To
Larry + Barbara;
How I spent my life's
working career. Jack

FROM GREASERS TO GANGSTERS

Four Decades of Delinquents

Jack L. Graves

Jack L. Graves

ISBN 0-9652434-1-9

Printed by Maverick Publications, Inc.
P.O. Box 5007 • Bend, Oregon 97708

AUTHOR'S NOTE

In my thirty-one years as a Camp Tillamook teacher five thousand boys attended my classes, and, in a few instances, three generations of family members. I have written of only fifty or about one in a hundred. Even though I could write a hundred books like this one, I chose to tell these stories because I found them interesting and hoped that others would too.

Teachers' memory banks become over-loaded after many years of taking roll and filing reports and report cards, so we are perhaps more likely to recall members of our first classes better than those of our last. Fortunately names weren't important so, except for Camp staff members, I changed those I did remember. Nothing I have written should be assumed to be factual although these stories are based upon my experiences teaching "delinquent" boys..

Prisons, especially, those for juveniles, can be somber places in which inmates engage in nefarious kinds of activities thwarted by staff only on a "catch as catch can" basis. If that were all I were able to write about this would be a sad report indeed! Over the years many people "on the outs" have asked what satisfaction I could derive from teaching "those kids." I hope these stories may help them to better understand the challenges, satisfactions, and disappointments, that we correctional teachers may experience.

Today, many jurists and penologists do not recognize youthful offenders as different from adult criminals and thus deserving of the same long sentences in prisons lacking in educational facilities or other kinds of self-improvement programs. They are wrong. Incarceration is opportunity; we should use that time well!

Jack L. Graves, Teacher
Camp Tillamook, 1957 to 1989

TABLE OF CONTENTS

THE GREASERS ERA

"Greaser" is the general description given to delinquents of the post-World War II era, the late '50's and most of the '60's. Greasers were loud, arrogant, and posturing. They were easily identifiable by their hair styles which were slicked (greased) and swept back on either side to meet at the back of the head or DA'ed ("like a duck's ass"). Elvis' hair styling marked the latter part of this youth culture era, as did tight legged (pegged) trousers, narrow leather and fabric belts, and short sleeved casual shirts, worn open almost to the navel.

Some of these youths formed territorial gangs (re: film "West Side Story") and worked on and drag raced fast cars (re: film "Rebel Without a Cause"). They prided themselves on toughness and despised school (re: film "Blackboard Jungle"). This era may be considered to have ended with the European (Beatles) influence giving way to the "stoners." ("Flower children" were older, idealistic collegiate youths who identified with the peace, love, and drug culture of the time and so were transitory between "greasers" and "stoners."

Most greasers who survived their teens and early twenties were absorbed into the greater society. This integration was aided by the draft's compulsory military service blending "greaserism" and military "machoism" into assembly line workers, husbands, fathers, and today's television-addicted couch potatoes.

CHAPTER ONE

THE RUGGY

Several youths were washing a bus and station wagon. Both Oregon green and marked with the State Seal. Another teenage boy stood nearby on a loading platform. A cigarette dangled from his lips bobbing up and down as he directed the others' efforts. "Hey Mike, get a step ladder outta the shed here an' wash the top." Because he appeared to be in charge I directed my question to that youth with a deferent timidity that surprised me.

"Excuse me, but could you tell me where I might find Mr. Rich?"

"I dunno. I ain't seen him since this morning."

As he spoke, I studied him. Charley was about six feet tall with very blonde hair which was meticulously styled in the popular Elvis-like manner, swept back on both sides to meet at the back of the head, "duck's ass-ed" it was called. "A handsome kid but surly," I thought.

"My name is Mr. Graves, I'm to meet Mr. Rich to have dinner with him."

"You gonna eat here?" he asked incredulously.

"Sure," I said. "Isn't the food good?"

"Cook's off today. House boys cook. Slop . . . jes' slop. Didn't Rich tell you. Tomorrow, come back tomorrow."

"Well thanks," I responded hesitantly, "but I'll risk it."

"Yeah," he said. "Don' say I didn't warn you."

"You work here," I inquired.

"Yeah," he said again. "My name's Charles Eggert, but my friends call me Charley."

"Well, thanks again, Charley," I said turning to leave.

"Charles, Charles W. Eggert. Okay?" he replied.

I didn't turn around or acknowledge that I had heard him then rounding the corner of a rudely shingled building I met Mr. Rich.

"Oh hello, Jack," he said. "Say I forgot that tonight is the cook's day off. The boys cook and while they do their best, but . . ."

"Charley told me. 'Slop' he said."

"It's not that bad. Mostly they just open cans and heat the contents. Still think you still want to stay?"

"I cook on a hot plate, and do the same in a smaller pot."

"Okay," he said. "Dinner will be in about ten minutes." He paused, then added, "So you've met Charles."

"He says only his friends call him Charley."

"I know," said Mr. Rich. "He doesn't have many."

The food was just as Rich had described. More interesting was the interplay between the inmates. They were loud and each vied for dominance in conversation, except for Charley. He seldom spoke, but when he did, there was a momentary hush. Others yielded to him. He took large helpings of the "slop", most of which he didn't eat. At the end of the meal, everyone took their plates to the garbage can and scraped them, placing their other tableware in shallow pans of soapy water. Charley and an older man whom Mr. Rich had introduced as a staff member didn't. They both gave theirs to a younger inmate to clean.

"Well," I thought, "they don't look any different from most kids." Still the idea of teaching all day in a public school classroom, then evenings in a 'reform' school didn't seem attractive but I wouldn't be under contract so I could quit if I didn't like it. . . . And the extra money would help pay off my car.

"How many boys are here?" I asked my host and possible future employer.

"Thirty, when we're full up, but we usually run closer to twenty-five. You know, they're coming and going. Most will stay here at the Camp for maybe four months . . . some longer, it all depends upon their commitment charges and how willing their home county judge is to have them back raising hell, but at least here they get paid a quarter an hour and learn to do a day's work. We send 'em to work in the state parks during the day up the highway at Ozwald West State Park and just across the sand dunes here on the Nehalem Spit where they're building a new State Park. We got several more years here, but as you probably noticed these old tar-paper shacks are on skids. Like the old logging camps we can go where the work is."

"What do the boys do here," I asked, "when they aren't working?"

"That's been a problem," Mr. Rich responded, "and part of the reason we want you. They've got too much free time. Television only goes so far, especially when we only get two fuzzy channels. Card games lead to gambling and fighting. This is designated as a full time work camp, but let's face it, there isn't a hell of a lot for kids to do work-wise when they leave here so their parole officers usually put 'em back in school then they're way behind. I'm hoping you will help us out—maybe three nights a week—a couple of hours each night. Whatever you can do will help."

I told Mr. Rich that I would give him a definite answer by the end of the week. He invited me to return for a "real meal cooked by a real cook." I did have a chance to chat with several youths to discover what their feelings would be concerning attendance in night classes. Opinions were divided but favorable if "the night classes don't include algebra or English, and if attendance ain't required." Several hoped, "it would be stuff that could help me get back into public school, like geography maybe." The last comment influenced me most. I enjoyed teaching geography and knew I could make it interesting.

Thursday I told Rich I would start the next week and gave him a list of text books and materials I would need.

The following Tuesday at six-thirty in the evening. I drove to the camp. Quite uncertain how I was going to start. The dining hall was the only room with adequate light and seating. Some youths were lounging about. Others were washing dishes and mopping the raw plank floors.

Charley leaned against the casement of the door obstructing my passage. "Hi teach," he greeted me sarcastically. His cigarette jerking about in clenched lips. "Boy, did Rich 'sucker' you!"

"What do you mean?" I asked fighting to suppress a sense of foreboding and thinking, "Maybe it's not too late to back out."

"Nobody here wants school."

"Oh," I paused, sensing he had more to say.

"Yeah, t.v.'s no good and card playing jes' causes fights, so you're it."

"I'm what?"

"Entertainment."

"Well, I don't sing and dance."

"Too bad," Charley said. "Maybe you tell jokes or do magic tricks . . . better do something exciting if you want to stay here!"

"No," I said. "I don't do anything exciting."

"Too bad," Charley said and sauntered arrogantly across the room to a chair which he tipped back against the wall as he sat.

Seeing Ray Rich across the room I shook his hand announcing, "I'm here," and with some inner reservation added, "ready to get started."

"Fine," he said. "I've got your class stuff here." He pointed to a small table on which were a box of pencils, a ream of lined filler paper, a small chalkboard and two pieces of chalk, and a single book with a red cover inscribed simply, *World Geography*.

"I forgot to get a chalkboard eraser," he said, "but you can use this." He said handing me a small piece of cloth.

I picked up the book. "I recognize this," I said. "I used it two years ago student teaching, it's a good book but I think each student should have one."

"One is all I could get," he replied. "The Principal of the school at Mac' said maybe I can get half a dozen more at the end of the term. Can you get any from the school district?"

"I'll check with the junior high. I might be able to borrow a few." I commented. "How many students will be in class?"

"Twenty-two," he answered.

"Everybody?" I asked.

"All, but three, who refused, so I'll put them to work cleaning the dormitory."

"Then class is compulsory?" I questioned.

"I can assign a staff to help keep order," he replied in an apologetic tone.

"No, let him supervise the dorm cleaners instead because if I'm going to take this job, I reserve the right to refuse to try to teach uncooperative or disruptive students. I'll send them to him to put to work."

"All right," he said, "if that's the way you want it."

"Let's get started," I said, surprised at how fast my pulse was racing.

While the inmates gathered I thumbed through the red textbook trying to decide how to start. My new students were surprisingly silent. They were sizing me up as I was them. Then in as strong a voice as I could manage, I said, "How many of you want to be here?" Half a dozen raised their hands. Several others muttered unintelligibly.

"Okay, those of you who want to go may leave now."

Nobody left, but every boy looked toward Mr. Rich where he stood at the back of the room with a staff and the cook. He nodded his head affirmatively, and four youths slid their chairs noisily away from their table and strode out.

"If you stay, I expect respect and cooperation. I'm here to help you learn . . . not to fight with any of you. Okay?"

One other student stood up and silently left.

Charley still leaned his chair against the wall with his feet on the lower chair rung. He inhaled deeply on his cigarette and blew a smoke ring which quickly became misshapen. He stared at me through its haze.

"Another thing, class will be two hours with a fifteen minute smoke break after fifty minutes. No one will smoke during class."

Charley casually stubbed out his cigarette on the recently scrubbed floor planks, fully aware that everyone in the room was watching. I wondered if he knew that I had just invited him to leave the room as the others had. When I realized that he had no intention of leaving I felt a momentary flutter of fear in the pit of my stomach.

"Well teach, let's get started. Okay?" Charley commanded.

"What do you want to study," I asked, holding up the red book. As expected some humorists shouted "sex" and "booze", but one realist finally asked what the choices were. Thumbing through the *World Geography* I read the chapter titles.

"The African Plain, the Antarctic Continent, Our Earth and Its Neighbors . . ."

"Hey my brother's at the south pole in the Navy. That's the Antarctic isn't it?"

"Yes," I replied. "Shall we start with that?"

No one said anything so I took this lack of response as an affirmative response and continued. "Since we only have one book I will start by reading the first three pages."

"The Antarctic continent is almost totally covered by ice. There are a few areas which remain ice free the year around because of volcanic vents and larger areas in which the bare earth is exposed during the brief summer months . . .," and so I read the first three pages. Looking up from my reading periodically was quite depressing since it showed that most of the youths seated at the dining tables were either asleep or about to become so.

I stopped reading and asked, "Anybody still awake?"

Charley, still tipped back against the wall and still awake, commented, "This is boring as hell."

"Well Charles, what do you think we can do about that?" I asked.

"I dunno'," he replied, "but I want a smoke."

"Okay why not? Wake everyone up and take a break."

Standing outside the door smoking were several boys as I stood just inside with Mr. Rich. I knew I had to do something different and fast. The several boys around the door were critical of what had so far transpired. One said he was bored as I read page after page; another wanted to talk about the Antarctic informally; while still another said he would prefer to answer the questions at the end of the chapter like they did in his home town school; yet another wanted to write a report.

After the break when my remaining seventeen students gathered I announced, "Each of you wants something different from this class and so we will try to have something for everybody. We will start by forming three committees of five or six. Each committee will then plan how and what they want to learn about the Antarctic."

Charley didn't move from his familiar position at the back wall, but it wasn't necessary. His committee formed around him. Charley was not really a participant although clearly in charge.

After an hour's discussion, satisfactory methods were devised and various materials were proposed for independent study. One committee formed into a discussion panel, a second chose to write reports. The third wanted to simply answer the questions posed in the book at the end of the chapters. John volunteered to write his brother to invite him to tell of his experiences in the Antarctic on his next leave. What had begun as impossible briefly seemed not only possible but exciting!

Several weeks passed rather uneventfully as we moved from the South Pole to the North Pole and then to the Kalihari which led us to the Sahara. Two holdouts joined our study group. As students the inmates were aggressively vocal, if woefully ignorant. Everyone had an opinion and wanted to express it at the same instant. That is, except Charley who sat silently, his chair always tipped against the wall, watching and listening. I seldom spoke to him and he had nothing to say to me. I resented his non-participating presence and decided to force him out when I found the opportunity . . . and the courage.

"Did you know it's against camp rules to lean or tip chairs against walls and sit on them?"

"No, I didn't." I responded to Rich's question. "You want me to make Charley untip his chair?"

"No, we don't cross him."

"Why is he treated differently from the others?" I asked.

"He's our 'ruggy'." Ray responded.

"What's a 'ruggy'?" I had never heard the term but it conjured unpleasant images in my mind.

"He keeps order in the camp," Ray said. "As long as he does it without disrupting routine. It's easier for us. Charley's good at it. He never leaves marks. . . ."

"What kind of marks?" I inquired.

"Cut lips, bruises, black eyes, worst of all, broken bones," Ray replied.

"You encourage this?"

"No, he came to us that way . . . Smith Cottage at MacLaren uses youth monitors which are of two species, 'ruggys' and 'rats' . . . and, well, Charley is . . . ah, useful."

"What is a rat?" I asked.

"A 'snitch', 'tattletale', 'rats' are despised while 'ruggys' are respected. If either varmint gets too far out of line some night after lights out he'll become the guest at a 'blanket party'. It's the law of our jungle. The little animals here can survive only by cooperation or craftiness."

". . . and what's a blanket party?" I inquired, even though I had already guessed his answer.

"That's as many as it takes to throw a blanket over someone and beat the 'crap' out of him. The blanket prevents the victim from identifying and later retaliating or revealing the identities of the party-goers."

"Does this happen often?"

"No," Rich answered. "Charley keeps it from happening here. He knows just how far to go with the boys and the staff. He's boss and will remain so until challenged or humiliated then he'll probably hike."

"You mean he'll run away?" I asked.

"Yes, and keep running until he finds someplace where he can regain his status."

"Who will challenge him," I asked.

"Not me," said Ray. "He's been a good ruggy, but someone will take his place when he is paroled or before that either by intimidation or by whipping him in a fight. Chances are our new ruggy will be worse."

"And what if there is no ruggy?" I asked.

"Chaos, turmoil, the inmates will challenge one another until one emerges dominant, fights and threats, some will openly rebel against us, and the system, to gain status. It's the 'shits' not having a ruggy."

Class had begun after break. It was quiet with most of the boys writing or reading. At the back of the room Charley sat tipped back in his chair.

"Charles, Mr. Rich tells me it's against camp rules to tip your chair back. Please sit in it so that all four legs are flat on the floor."

"Why? I've always sat this way. You've ain't said nothing before," Charley responded bellicosely.

"I didn't know it was against the rules. Besides it can break the chair, damage the wall, and it's dangerous. If your chair slipped you could hit your head or something."

Charley feigned a look of astonishment replying, "You could get your head or something broke too. Maybe by this chair."

"Are you threatening me?" I asked. Now that the show-down had come, I was ready to bring it to a head, no matter the consequences.

Charley glared but was silent.

"Talk's cheap, put-up or shut-up." I said, with bravado for I knew Charley was too smart to risk the consequences he would face if he assaulted me.

He stood up and glared, his face was twisted in an angry mask of animosity. He lunged forward, stopping just short of bumping me with his body. Then wordlessly strode out the door. The room erupted into excited comment.

"That's over now. Everyone can get back to what you were doing," I told the class.

"It ain't over," one youth said grinning.

He was right. Things weren't the same without Charley. There was constant bickering and insulting remarks, calling one another stupid, several times coming close to blows. Six students left the class as we dropped to eleven students.

As for Charley, I saw him sometimes during smoke breaks. When he came across the yard to join our group, conversation ceased. The boys smoked in silent anticipation. Charley was sullen, casting evil stares at me as he attempted to maintain his reputation among his peers and preen his self-image.

One night just before Christmas, the evening was cold, crisp and clear. One of the boys pointed to the darkness of the sky and said, "Look, a falling star."

"It's not a star you jerk! It's Sputnik, the Ruskie's earth satellite," said Charley. "It goes around the earth in an hour and a half and radios a beeping signal so's it can be tracked. That way because of a thing called doppler they can tell its speed and how much it's slowing its speed."

"What keeps it going?" asked one youth.

"Inertia" answered Charley.

Charley was talking about the doppler effect and inertia! I was too much amazed to speak, and it was well that I didn't.

"What's that?" asked one.

"It means when it gets goin' it jes keeps goin' cause there ain't nothin' to stop it."

"Yeah?" another said skeptically. "So how's it get back?"

"It don't but someday years from now it will fall down to earth."

"What if it hits somebody on the head?" asked another.

"Can't," replied Charley. "It'll burn up first."

"Why . . ."

I interrupted. "Time to go back to class. Hey Charley, why don't you come in and answer all these questions about Sputnik?"

He paused, but then said, "Okay."

Later that month I wrote my first case reviews. I wrote "Charles W. Eggert" at the top of the form, "age 17." I thought for a long time and then wrote: "Charles compensates for his feelings of intellectual inferiority by arrogance and anger. He needs to be rigidly held to the same rules as other boys in the camp-setting. Staff should reward his good conduct with praise and accordingly place trust in him as he earns it. Charley wants to return to school at home."

In mid-March it became apparent that two teaching positions were more than I could handle. I decided to give up the night classes at Camp. Charley had become an enthusiastic and confident student, a bit overbearing perhaps, you might say an "intellectual ruggy." He was my unofficial teacher's aide and ex-officio sergeant-at-arms . . . but he sat in an untipped chair.

My treatment plan which seemed so simple and obvious to me was widely circulated and when it was brought to the notice of the Superintendent was responsible for my employment in another camp that next school year and thirty years more beyond. As for Charley, I never heard of him again, but as I was to learn, in the delinquency business, that's success!

CHAPTER TWO

CAMP TILLAMOOK

Charley's reputation created mine. This seeming success story had reached Amos Reed, Superintendent of MacLaren. Mr. Reed had come over twice to talk to me when I was still teaching at the Camp Neacarney, but since I had resigned two months earlier, I didn't expect to hear from him again.

He was determined that I should become the teacher at a new camp which had opened in an old barracks on the abandoned naval air station in Tillamook. There had been problems at Camp Tillamook which resulted in the resignations of two staff and the firing of another so there was immediate need for a teacher. I was looking toward a long and restful summer of leisurely travel and recreation which is after all one of the few benefits enjoyed by classroom teachers, but after several telephone conversations and a telegram I agreed to come to Woodburn to "at least talk about accepting the teaching position at Camp Tillamook."

So I drove to Woodburn less than a week after beginning summer vacation to interview and be interviewed. I was astonished by the number of interviewers I faced. We started with six and then were joined by three more.

Mr. Reed had already told the others of my work at Camp Neacarney and especially about Charley. They asked me many questions as to how I would handle hypothetical situations involving youth conflicts and crises. The school principal came in and we were introduced. To my questions, I was told Camp

Tillamook was a new kind of program intended to provide work experience without sacrificing education. This new camp would provide half-day school and half-day work to thirty boys who were all to be academically advanced at least high school juniors and seniors. I would teach all subjects but the sciences and mathematics which were to be taught evenings by two local teachers employed part time as I had been at Camp Neacarney.

When I hesitated. Mr. Reed asked if I would be willing to work for just three weeks in the place of a line staff who was going to the National Guard's summer training camp. After those three weeks of work as a line staff I could leave if I wished or stay as the camp's full-time teacher.

I agreed. Three weeks wasn't too long a cut from my summer plans. Mr. Reed had embarked on a massive reorganization of MacLaren, and I was included. Certified teachers who were qualified from grade six through high school were difficult to find at the salary offered. The offer of full room and board for only fifteen dollars a month more than compensated for the lower salary. I might have refused Reed's offer had I known that I was to have to postpone my vacation plans for three years.

The following Saturday, I walked into the old barracks building a half an hour early. My initial impression was of cleanliness. The deep red linoleum shone from buffing, and the windows, walls, and furnishings, while old, glistened with furniture polish and new paint.

Such cleanliness was not by accident, Saturday is clean-up day in institutions the world over. The old barrack's entry was wide with two flights of stairs connected by a bannistered landing led to a second floor. The entry area, held a reach-in soft drink vending machine of the sort in which bottles were suspended vertically on parallel metal rails in chilled water. There was an old, rather saggy upholstered sofa with newly varnished maple arms, and white porcelain drinking fountain, as well as two coin operated candy vending machines. On a small end table was a shallow china dish containing several coins and a dollar bill which I assumed was there to make change for candy or pop purchases.

A man perhaps ten years older than I introduced himself as Kelly then directed me to the door directly across from the main entry's double doors. It was a large room in which a man of about fifty sat behind a large black topped desk.

He rose to meet me and extending his hand said, "Hello, I'm Claire McMillan, Camp Director, and you've just met Kelly Palmer who is our weekend staff. You'll be working with Kelly today and tomorrow. He knows the ropes and will break you in."

"I'm Jack Graves." I said, quite aware that my name was known still it seemed unnecessary to identify myself, "hired to work in place of one of your staff for three weeks. After that, as camp teacher . . . if it works out."

"Yes," he replied, "Mr. Reed told me about you and I called Ray Rich. They both spoke very highly of your work at Camp Neacarney."

"Well, I'm glad to hear that," I replied. "I think I will like teaching here better. I understand the boys in this camp are selected for their scholarship and that they are minor offenders."

"Yes, most are," he replied. "We have five boys from Albany who were committed on 'minor in possession' charges. Their judge was furious because they wouldn't tell where they got the booze so he sent all five to us. Mr. Reed felt they would do well here. Then we have a seventeen-year-old student from Portland who has stolen sixty-three cars. His parents bailed him out again and again, until they couldn't any more. He has a three after his name. We call him Richard the Third. He may find that funnier after you teach him some Shakespeare. We also have two sixteen-year-old Eskimo kids. Oregon contracts with Alaska to house their offenders since there's no training school there yet. Neither boy can read or write much, but then we don't expect them to. They're seal hunters not poets. We also have kids from the Warm Springs reservation, a Klamath and a Umatilla. We have boys from all over the state, a big friendly kid from St. Paul, several from Salem, and a tall colored youth from Portland who talks constantly. We have twenty-six boys just now. By the end of the week you'll know them all."

"I was told that these are hand-picked students," I said.

"We have three seniors now with six more waiting to come," he replied, "and the Eskimo and Indian kids always do their best. They are good boys who just drank too much and raised a little hell. It will be three weeks before you start classes. By then you'll see what I mean. Well, I have to leave. I just came out today to meet you. Kelly and you will take the group on town visit in a few minutes. See you Monday. Goodbye and good luck." He shook my hand again as he left.

Kelly flipped a switch outside the office door marked with a red stencil-painted sign reading "FIRE" a very loud bell sounded and boys began to gather in the entry. They were neatly dressed and clean, ready for their shopping excursion. As Mr. McMillan had said several races were represented.

"Two dollars each. Every boy gets the same amount of spending money which is deducted from his account," he explained. "It's money they have earned working here on the base. They do clean-up and maintenance and are paid twenty-five cents an hour for four hours work. The other four hours they are in school."

The boys filed onto the bus. They glanced in my direction curiously. Kelly and I stood at the front of the bus. He said, "Listen guys! This is Mr. Graves who will he taking Mr. GrosJacques place for three weeks then he will be your new teacher." Kelly looked to me expectantly.

I said, "I'm glad to meet you all, and I'm sure we will become much better acquainted."

Surprisingly no one commented. The trip into town was uneventful. The only youths I saw in town were loudly teasing two girls on the street. With my best teacher's scowl and sharpest reprimand I could muster, I sent them on their way.

The girls didn't seem very appreciative. One muttered something like, "What's it to you?" and the other made circular motions around her right ear which obviously questioned my sanity. I put it off as all in the day's work, but I understood their displeasure when those three boys did not return on bus and so I had to conclude that I had driven off the girls' boy friends. Kelly and I stopped for coffee and when we returned to the bus everyone was there waiting. "Easy money", I thought.

Saturday evening was uneventful. The group knew what to do and were settled into the routine. I had to check the schedule on the desk frequently to find out what was to happen when. That evening I left, but I would return the following morning and bring with me an overnight bag for I would stay over Sunday night.

When I came back Sunday everybody, except for five Roman Catholic kids who were dropped off for mass, went to the same Protestant church after which we ate an abundant ranch style dinner. Then both Kelly and the cook left. I was on my own, but by now, thanks to my earlier successes, supremely confident.

The inmates spent the evening watching television, playing cards and wandering about talking. No one talked very much to me but those who did curiously asked about where I lived, where and how long I had taught, and several, had I ever been arrested?

After referring to the daily schedule under the glass plate which covered the desk top, I started the group to bed at nine. "Nine p.m.—bedtime, ten p.m.—lights out." The boys slept on bunk beds, four to a room. Each had half a tall stand-up wardrobe cabinet and half a walk-in closet for storage space. There were no doors on the rooms. As the boys were getting ready for bed, I strolled the halls enabling me to visually observe preparations for bed. By twenty minutes after nine all were in bed. Some were reading and others talking, but most were just lying awake, I was pleased that things were going well and that the group were in bed so quickly.

"You can't have your light off," I said, reaching inside one darkened room and switching the light back on.

"Why not?" asked one of the occupants.

Another added, "Yeah, why not?"

"Schedule says lights out at ten," I replied.

"Screw the schedule," a voice said from the room across the hall. "I can't sleep with the light shining in my eyes."

"Read or something," I said.

Down at the end of the hall a commotion erupted as a pillow flew across the hall. The noise level had increased rapidly. Obscene shouts erupted in another room. Running into that room I found two youths toe-to-toe with clenched fists. In an adjoining room the sounds of breaking glass mingled with laughter and curses as shoes, pillows, and half-naked bodies darted from room to room.

Turning off the lights I muttered, "Screw the schedule!"

Claire D. McMillan, Camp Tillamook's first director.
The original camp building is in the background.

CHAPTER THREE

THE ANGRY ONE

"Jack, how would you like to coach baseball?" McMillan asked.

It was my second week as a camp staff. There had been no mention which I could recall of coaching duties in my discussions with Mr. Reed or Lord High School's Principal, Warne Empey.

"Well, I've never coached baseball. I have coached basketball, but baseball is a sport that I know very little about. Where's the equipment and what other teams could we play?"

"We were given uniforms and equipment from Tongue Point Naval Base at Astoria. It's an odd combination of softball and baseball stuff. What we don't have we could get from MacLaren's recreational department. There's an industrial league we could pick up games with this season, and probably join it next season. But even if we didn't play a single game practice is something the kids could do evenings. We can always make up games like staff against boys or inter-team games. The cottages at MacLaren each have a team and play each other. We could drive over a couple of times if necessary to play some of them."

"How many of the kids want to play. Do you know?" I asked.

"Logan Redbird and four other boys asked me about it, so there's at least five. I'll ask for team volunteers at dinner tonight if you'll agree to coach the team."

"Only if twelve or more are interested in playing, then I'll try," I said.

That evening at the meal there were eighteen who said they would like to play. I scheduled our first practice for four p.m. the next day. That morning I went to the public library and checked out a well-worn book entitled *Baseball: From Sand Lot to Pro*. I read some of it during the lunch break and as much more as I could that afternoon before practice.

With the debacle of my first night in the dormitory fresh in my mind I was determined not to rush into anything or to follow any schedule other than my own. My library book taught me how little I knew about baseball so I determined to start slowly with warm-up calisthenics and short throwing practice the first evening to prevent "sore arms". The next evening I demonstrated infield and outfield positioning followed by infield throwing practice and base running. I had cut the fat part of a wooden bat lengthwise to make a fungoe bat and hit pop-ups and grounders. I knew that I could control these activities simply by withholding such equipment as catcher's gear and bats.

The equipment was as Mr. McMillan described. The uniforms were really not for baseball, but softball. We had pants, socks, and t-shirts, but no caps. There were a great number of baseball spiked shoes, some seemed hardly used. There were plenty of gloves, both baseball and softball, but only one first baseman's glove and one catcher's mitt, two catcher's masks and shin guards, a chest protector for the catcher and one for an umpire, two dozen new balls and that many more used ones and thirty oddly varied bats.

Very few of the camp boys had ever played organized baseball. Little League teams were just beginning in 1958. A couple had played in Junior American Legion leagues and one or two more had been on their high school teams. Mostly we were all rookies.

Our field just behind our building had recently been used to pasture heifers and so was covered with cow manure flops and hoof marks. The grass was tufted in clumps and it was very uneven with first base on a mound and third base in a hole. We eventually moved our first base mound for the pitcher to pitch from, and scalped enough from the infield to fill the third base area. By importing several pick-up loads of fresh clean sawdust from a nearby sawmill we made a fair infield and base paths, at least better than the ruts they had been. With some two-by-four studs and chicken wire we constructed a backstop of sorts.

Our first day of practice began my second week as a Camp Tillamook staff. I was still learning the names of some of the boys. We started by forming two double lines thirty feet apart and throwing baseballs back and forth. That was short-lived when boredom set in and games of "burn-out" began with partners throwing at each other as hard as they could. Bored by that they then began flinging cow flops at each other. I cut practice short.

The next evening was equally disorganized. When I saw horseplay begin I brought out a fungo bat and after hitting flys and pop-ups for fifteen minutes I was forced to begin an impromptu game of "21" in which I hit to many different players so that no one would "get up" before practice ended.

When I finally brought out the bats and catcher's gear everyone crowded about me volunteering to be pitchers and hitters. No one wanted to field so I spread boys out so I could hit pop-ups and grounders for them to field then to throw to the base called out. Every other swing, I would hit to the outfielders who were to relay their throws through a cut-off man to first or to the catcher at home plate. Since we hadn't built our backstop yet I positioned two players to back up the catcher. I was hitting several balls one after another to speed the action. This meant that at any moment two to four baseballs were in flight.

These balls began reappearing suddenly in close proximity to my head. Some were thrown from the infield and others from behind me where two players were shagging balls. I was caught in the middle making it impossible to discover who was throwing baseballs at me and who was returning them to play. Concentrating on coaching while avoiding being hit was at best difficult. After the first few near misses, I was foolish enough to believe my loyal team members were throwing near me by accident, but it soon became obvious that I was escaping thanks to their inability to throw more accurately.

Since survival had become a matter of luck and alertness, I switched to an activity which did not put me on the firing line. By the time our first week of practice ended I knew I had to do something to assert my authority or quit. I had no idea what to do. I was desperately angry.

That evening I again called an end to the practice, sooner than planned and was standing on the stair landing halfway between the first and second floors. Only a waist high railing separated me from the first floor six feet below. Two youths whom I had asked to bring in the equipment were placing it in a small closet beneath the landing where I stood unnoticed. One was a blocky, muscular youth named Jones and the other Logan Redbird, a sixteen-year-old Indian youth.

"I almost hit that jerk coach a couple times. Tomorrow I'll nail him for sure."

I was instantly angry. Jones had thrown at me! Without thinking I vaulted over the railing and landed astride his neck and shoulders smashing him into the floor. I seized him by his collar, yanked him to his feet, and threw him backward across the entry hall into the office where Mr. McMillan sat working at his desk. Jones fell in a heap on the office floor in front of his desk.

McMillan looked at the sprawled body and then to me. "I've been wondering how long it would take you to get angry enough to do that. I've been watching your practice from my office window. I assume that you are referring Mr. Jones to the office."

I was livid with anger. "Yeah, he tried to hit me with a baseball during practice. I won't put up with that!"

"I don't think he will do it again. Will you Jones?" McMillan said with a faint smile. Jones mumbled something. McMillan stood up and shouted across the desk at the hapless Jones, "Now get out of here! The next time Mr. Graves brings you to the office, you're gone, on your way back to Mac'. You understand me!" McMillan screamed.

Jones backed out. McMillan looked at me and said very calmly, "Welcome to our staff, coach."

I was trembling; adrenaline rushed through my body. My voice sounded shrill and seemed to crack as I said, "Aren't you really angry? You were just screaming at Jones, now you don't seem mad at all."

"It's part of the job. You know the best defense is a good offense. Word will spread. No one will ever throw a ball at you again. They just wanted to try you to see what you would do. Now they know," he smiled, and added, "and we're going to have a baseball team."

He was right. No one challenged me, or threw at me again. Jones became our third baseman and Logan Redbird played second base. He had grace and coordination, a good arm, and a sixth sense which enabled him to pick off runners who took too big a lead from bases. He seldom threw the ball away. Unfortunately not so the rest of our team! We made too many fielding errors, but lost all our games at the plate and from the pitchers' mound.

Logan and I became well acquainted. His family had traveled the Northwest and Rocky Mountain states as far east as the Dakotas. They lived on the Rosebud Reservation, and the Wind River Reservation, and went to Pendleton for the Roundup. His brother had been killed in a shootout with a State Police officer. Logan had a great deal of anger about this since he said his brother was drunk and unable to defend himself when the policeman started shooting.

His family were Shoshone, Ute, and Nez Perce. They traveled because of their involvement in Indian tribal affairs, and were considered prominent tribal leaders among all these tribes by birthright. Logan had been committed for assaulting and threatening to kill the officer who had shot his brother. His anger was such that I was sure that in the end it would destroy him and those around him. However, my recollection of Logan is that of the graceful second baseman who could be depended upon to cleanly field the ball and throw out the runner.

Making the best of cooking outdoors after the Columbus Day Storm, 1963. Mrs. Nina V. Palmer, the first camp cook from 1957-1969.

CHAPTER FOUR

PUBLIC ENEMY #1

"Hey George, the girl you was neckin' with at the movie on Saturday night is a judge's daughter. I betcha din't know that did'ya?" Frank shouted across the room to George Hurley as they were shaving.

Both boys, bare to their waists, were in the bloom of youth. Frank Cottra was swarthy with jet black hair and an aquiline nose, while George who was not as tall was freckled and wore his straw colored hair in a flat top. George was considerably heavier than Frank and several crudely etched tattoos contrasted sharply with his pale, almost albino flesh.

"Sylvia is a judge's daughter?" George's retort was as much an exclamation as question.

"Yeah, that's what Mr. Aho told me," replied Frank dabbing at his newly shaven face with a small white terry towel. "You know how many years you can get for knockin' up a judge's daughter?" Frank laughed.

"No, tell me how many, lawyer man!"

"Life, with a shotgun salute," Frank replied as he took a swig of red liquid from a bottle and gargled loudly.

"I don't care if Sylvia is the President's daughter, an' besides ain't no chance o' makin' out in a theater full'a people. Could you imagine gettin' romantic in a movie like 'Beast from the Black Lagoon'? No way that's not the kinda movie that turns on girls, maybe an Elvis flick where he sings 'Love Me

Tender, Love me True' and wiggles his ass around in tight ass bell bottoms. Yeah, that's the kinda movie girls like?"

"No, no, Daddy'o, here's what they really want," Frank pointed at his smooth olive colored torso, "A real man like me an' I gotta hair on my chest to prove it! Do you have hair on your chest?"

"Lemme see your hair," George moved next Frank and squinted at his chest. Then he made a circle with his fingers and looked through it. "Yeah, there is a hair there, just one. Takes a magnifyin' glass to find it though."

"Hey guys!" George shouted to two others who had been combing their hair before the mirrors over the long row of white sinks. "See this, Frank's got a hair on his chest. Sammy, do you have a hair on your chest? No, I'll bet you don't, or you, Henry?" George addressed his last question to a very small immature red haired youth.

"I know it ain't fair when some of us got sex appeal and others don't but that's life!" Frank laughed.

By now attracted by the commotion several other boys arrived.

"Hey guys, look at Frank's hair. Do you think it's right he should have a hair on his chest and not share it?"

"No, George!" three or four boys shouted in unison.

"Let's take it an' give it to Henry," George shouted, advancing on Frank with mock giant steps. What had now become a merry mob fell in behind him mocking his giant steps.

Frank retreated into a six-person shower stall but with a rush they were on him. George stepped back to permit his cohorts to go around him then watched the struggle in the shower.

A voice cried from the darkness of the shower, "I've got it! I've got it!" and Pat, a skinny kid with thick glasses, came running out holding his invisible prize in the air.

"Give it to Henry!" someone shouted.

"Yeah, he needs it! Frank can always grow another."

"Okay," George said, "Jake, you go get a piece of Scotch tape," and one youth left the room on a run.

"Shouldn't we have a ceremony or something?" another asked.

"Naw, I wanna play a game of pool before dinner," George said.

"Here's the tape," the breathless youth said re-entering.

"Here ya are, Henry," said George, handing the other boy a tape dispenser and the invisible, but highly prized hair.

Henry stood holding his prizes. "What'ya want me to do with these?" he asked.

"Stick it on yur pore skinny chest," one of the Eastern Oregon boys referred to by the Portlanders as "'billies" advised.

"Yeah," George said, adding, "We'll be looking inside your shirt every day for it so show it off."

The group broke up into twos and threes, dispersing rapidly in several directions. I wandered down the hall behind George and Frank. Fortunately they were laughing, for such incidents all too often resulted in anger, but this time had been viewed by both the victim and the victimizers with good humor. Probably only two physically matched youths like Frank and George would not have fought over such humiliation but neither wanted to test the other.

As George entered the room where the pool table was two boys left the table and placed their cues in the rack. George knew he was supposed to wait his turn to play, but, as he explained, "If players just happen to quit when I arrive that's okay, right?" He and everyone else knew their quitting was not coincidental.

George was a bully boy. In some institutions he would have been the staff "ruggy", but we were trying to discourage his self-serving bullying and braggadocio. He wasn't a bad kid, although he had a lengthy criminal record, mostly for car theft. Joy riding. George said he only "borrowed" these cars. In the beginning from family, friends, and neighbors, but when he began running out of friends and neighbors he started taking them from downtown parking garages, until finally his parents however influential could no longer save him from legal retribution.

George liked to think he was worse than he was. In fact, his commitment charge was for jaywalking. In the days before Gault vs. Arizona juveniles were treated differently from adult criminals in that they were most often charged with the most minor offense the judge could find on their rap-sheets, usually a misdemeanor. In George's case when an officer started giving him the standard "Why you shouldn't jaywalk" lecture, George retaliated with grossly obscene disrespectful verbosity, which in turn caused a judge to declare him "in violation of his probation" and to order him sent to MacLaren. His commitment charge simply read, "Violation of Probation, jaywalking." So it was that George came to us as a convicted "dyed in the wool" jaywalker.

"Hey, George, What'cha goin' to do this home visit?" a youth at the end of our table asked.

"Probably swipe a car, pick up some babes and joy ride," George replied. "I can cross wire a car so fast it'd make yur head spin."

"Oh yeah, I heard about you 'round the big town. You was noted for drivin' nice cars," another kid said.

George rose to the challenge, "Yeah, T-Birds and Corvettes were my specialty, when I could find 'em. One night I outrun the cops on Sandy Boulevard in a Corvette doin' a hundred and ten."

"Wow! That's neat," a younger boy commented. "What kinda cars was they drivin'?"

"This conversation is turning pretty negative," I commented. "Why don't we talk about something more wholesome—like maybe sports, movies, or television."

"Fords, I think, I was so far ahead of 'em I couldn't tell."

"Sounds like you must have led a really exciting life before coming here, right, George?" Mr. Nelson commented.

"Oh my God," I thought,"now he's got the staff joining his fan club as well as all the pimply faced kids in camp."

"Yeah," said George. "There wasn't any cop car could catch me!"

"So how'd you get arrested?" Keith asked.

"On foot," George replied.

"Then you don't run as fast as you drive?"

"I told you, behind the wheel I can't be caught!"

"But, if you were caught on foot how could you be . . .?"

"Hey man, I'm gettin' tired of answering all these stupid questions. They hauled out my old rap sheet, okay?"

George's admirers at the table were beginning to sense discomfort.

"Yeah, hey, George, how could they arrest ya fer car stealing if you ain't in a car?" asked one of the smaller and younger boys.

"Easy, you jerk," George replied, "for probation violation. Now let's talk about something wholesome like Mr. Graves said. How's about President Eisenhower's ailments. Hey Mr. Graves, is that tendonitus or phlebitus Ike's got?"

I was ready to change the subject like George wanted, even to Ike's medical ailments if necessary, but not Keith.

"Then what is your commitment charge George?"

"Whad'ya mean?" George inquired.

The youths at the table were now following their conversation much like a tennis match. They knew George was being evasive, but didn't know why. Then I knew what Keith was forcing George to admit.

"Well, George, what were you sent here for? What terrible, awful crime finally got you arrested? Out with it! You won't shock us 'cause we all know what a bad ass you are!"

George paused for a long time, then looking helplessly from one of us to another answered, "Jaywalking."

There was a stunned silence, then one by one the boys at the table began to snicker.

***

I never again heard George brag of his criminal activities. Several years later I heard that he took a wife, a former judge's daughter, and has two children. For many years he has operated a successful auto repair business in the Portland area.

Russell Curnutt, Camp Tillamook's science and math teacher from 1961-1965.

CHAPTER FIVE

ZORRO'S SHINY WHITE STEED

Louie made an immediate impression upon me, not by what he did, but by what he didn't do. Louie did not like school and did not intend to do any class work.

Louie Littlecub was an Indian youth committed from Portland. His family were notorious. Three girls, four boys, and mother and father were all incarcerated at the time Louie was in my classroom. They were at most of the state's criminal facilities. The youngest was being held in detention awaiting his twelfth birthday at which time he could be legally transferred to MacLaren.

Louie's father was a huge man who was on temporary leave from jail when Louie and I met him on a field trip to the Oregon Centennial Exposition in 1959. He was a violent man known to be dangerous when drunk. Louie's mother and an older brother were serving time for separate murders. Two sisters were in juvenile custody and an older sister in prison. The brother nearest to Louie's age was also at MacLaren.

Louie seemed salvageable. He loved horses. His family had come to Oregon from Montana, and his dream was to return to the open grasslands and horses that he remembered from better times. In the late nineteen fifties the tribes were changing. Indians were traditionally seen as alcoholic and shiftless, but some tribes were building industrial and recreational facilities on their reservations. Other less visionary tribes had sold their lands to timber companies. Indian youths were caught and confused. They didn't understand. Some had too much money. Others, like Louie were dirt poor "dog eaters." The wiser tribal councils budgeted riches and urged patience to the impoverished and largely uneducated families in their jurisdictions. Louie's family didn't get any of this "Indian" money and didn't expect any for they were tribeless nomads.

His first day in class I awakened Louie repeatedly, once in frustration yanking him from his seat. On day two I asked him to read aloud a passage from our tenth grade history book. Surprisingly he did try, but it was obvious that his reading level was deficient. Without his cooperation to test I guessed it to be about fifth grade level. Louie stopped reading and announced that he would never read again. He didn't, but in return for acceptable daily attitude reports he agreed to follow along word-by-word as others read. This kept him awake. I checked to see that Louie dragged his forefinger along each line as others read it. Louie, like the other Indian boys that I had had in my classes could be relied on to keep his promise. He wrote frequent letters to various family members in penal institutions. I helped him write the letters and so became well acquainted with those family member's problems and I was at the same time able to teach Louie not only letter writing, but also spelling, and rudimentary grammar.

After two months of his stubbornness, I recommended that he be returned to MacLaren as a Camp failure. The Case Review Committee refused to follow my recommendation because previous reports showed that Louie had refused to attend school there, and had been involved in many instances of disruptive conduct. I felt betrayed since I had been promised "hand-picked" academically superior students. But at least Camp Louie wasn't disruptive . . . just disappointing.

As the months went by Louie and I became "pen pals" as we wrote to his family. He always kept his finger moving along the lines in the text books while others read. He listened to class discussion and lecture sessions intently so I assumed he must be learning something and continued to report him as a well-behaved if non-productive student, just good enough to qualify for full privileges.

In ninth grade literature classes we read *Silas Marner*, Sophomores were doing *The Ancient Mariner*, some short stories by Poe, Thurber, and O'Henry, and in Junior English we were just beginning *Julius Caesar*. Since I had grades nine through twelve in the same room, I didn't object if students elected to put aside their class work and participate with another level. I laughingly referred to this as the Camp Tillamook liberal arts program. Louie liked to listen to the stories and for months dragged his finger with most levels of literature class.

"Mr. Graves, I want to be Cassius today," Louie announced one day as we were continuing our reading of *Julius Caesar*.

I was astounded. "You want to read the part of Cassius?"

"Yeah," he said. "I like Cassius. He thinks like me."

It was a comparison of personalities so farfetched that I could never have imagined it, but I agreed to let Louie read the part of Cassius as he argued with Brutus before the battle at Phillippi. I told Louie that if he didn't know a word to raise his hand and I would help him pronounce it. He only raised his hand twice but mispronounced a few words as we all did when reading the unfamiliar language of Shakespeare's day.

This was the breakthrough! Louie began doing his class work, as well as commenting and participating in everyone else's. Tests, which he now agreed to take, showed his reading level to be close to his grade level. He particularly liked European history in the days of the knights "because they rode horses." He spent his free time watching television and reading Zane Grey's novels. Louie's future began to look promising.

Then came crisis, the Mark of Zorro is a boldly slashed "Z." Zorro was running amuck in the camp, slashing "Z's" on bedroom and bathroom walls. The staff were under orders to capture Zorro. Pocket knives were seized, interrogations, and secret sources were utilized. Still Zorro struck time and time again. Everyone was at wit's end. I should have suspected

Louie. Zorro always slashes his mark from the back of a beautiful rearing black stallion. Louie had sufficient imagination and love of horses to turn even a porcelain toilet into the rearing mount from which a "Zorro" could scribe his mark of disdain to the State's authority over him.

"It's Louie," Mr. McMillan said. "Louie is Zorro. I've confronted him, but he won't admit it. He won't agree to stop. He's got to be returned to MacLaren."

"How do you know it's Louie?" I asked, knowing in my heart that Zorro and Louie were one.

"Staff caught him coming out of a toilet stall with this," he said holding out a pointed can punch. "There were 'Z's' in that stall. I want you to write a special incident report to return him for failure to do his school work."

"I can't do that," I responded. "The last five weeks he has been doing very well."

"He'll run away unless we send him back now," McMillan said. "If he would just confess. We'd have him paint a couple of toilet stalls and take him off full privs' for a week, but he won't give in, and I'm sure that he won't stop."

"I'll talk to him," I said.

I did. I used every approach I could. He wouldn't admit guilt. He wouldn't promise that no more "Z's" would appear on the walls. He was stoic, his only words were, "Send me back to Mac' if you want cause I'm not sayin' nothin'."

The next morning I opened my door to the hallway I shared with the inmates' rooms. A scrap of paper lay at my feet.

Scrawled in a familiar hand were these words, "Mr. Graves I am running away tonight do not think it is cause I hate you cause I dont," and with big sweeping "L's", very reminiscent of "Z's", he had signed it, "Love, Louie."

***

The last time I saw Louie in person was two years later. He came back to visit Camp Tillamook driving a shiny new red Chevy convertible. I was impressed, but not surprised since several Indian kids had come back driving new cars purchased at the price of their birth-rights. As I watched him drive away. I recalled that he was not affiliated with an Oregon tribe or reservation so I immediately reported the description of, and the license number of the auto. In less than an hour the Sheriff's office called back thanking me. It seems that Louie had run away from MacLaren that same morning, stolen his showy steed, and come to visit his old friends. I was sorry I had to turn him in, but it was probably for the best. Since that day I've only seen him on television news as an advocate for native-American rights to their sweat lodge religious practices at the Oregon State Penitentiary where he has spent most of his adult life.

My classroom desk and file cabinets, with the main office in the background—across the hall was the day room, the main door was to the right of the office, and the flats door to its left.

CHAPTER SIX

NO MORE NAME CALLIN'

Little did I imagine when I met Stanley Byman in 1959 that he would prove a pivotal factor in my decision to retire thirty years later. Brown vs. Topeka, the historic Supreme Court case, had heralded a new day of racial equality only five years before when it had declared racial inequality in public education illegal. Stanley Byman hastened the dawn of racial understanding at Camp Tillamook in his own special way.

Stanley's demeanor was aloof, but not arrogant. He was friendly to everyone and was a model of cooperation. He was conscientious in school even though his Wechler indicated an IQ in the dull-normal range of 87. I liked him and he seemed to like me. I was not discouraged with his ability level because Stanley was willing, even eager to learn. At the time of the incident, he had been at camp almost three weeks and was making excellent progress in reading and ninth grade geography.

It was against state law and camp policy to permit inmates to smoke, yet nearly all did. By a small creek a hundred yards from the camp building was a shooters' shed on skids which had been abandoned by the Rod and Gun Club and dragged there to serve our boys for unsanctioned smoking and, of course, crude socialization. Language and conduct deemed unwholesome were prohibited in the camp building, but since staff were not present around the "smoke shack" youths visited it at their own risk.

"Pressurizing" is a term unique to juvenile and adult correctional facilities which means that one inmate may force his will upon one or more others by threat or physical coercement. We had been having a lot of trouble with several larger youths who were dominating, even terrorizing smaller kids. "Pressurizing" those younger youths to give them cigarettes and to run and fetch for them. Stanley Byman was one of the smaller kids in the camp, and was black as well, a natural target for bully boys.

I was sitting in the office smoking, a privilege accorded staff, but not inmates. The students had gone outside to the smoke shack for a fifteen minute break. The office door was directly across the entry hall from the main door so when a boy came in that door dripping blood from his recently bloodied nose, I called for him to come into the office.

"What happened to your nose?" I asked.

"Jes' messin' around, an' I was accidental hit," he said.

"Well, you better go to the flats sinks and put cold packs on the bridge of your nose, and try pinching it like this," I demonstrated. "That may help. If it doesn't, come on back, and I'll pack it for you."

He had just disappeared down the hall to the rest room which inmates of correctional facilities call "flats", when another youth came angrily storming into the office.

"That nigger son-of-bitch hit me!"

"Oh, yeah," I said. "You must mean Stanley?"

"Yeah," Henry replied. "You tell him he just better keep his hands to himself or I'll kick the shit out of him!"

"He just hit you without reason? What caused him to do it?"

"I dunno."

Then we were joined by Ralph. A big red haired kid with a recently split lip. Next was George, wearing a new bruise on his cheek, and then Pete with a rapidly swelling eye.

Soon I had an office full of irate victims of Stanley's fists.

It was time for school to resume when Stanley strolled in the door as if nothing unusual had happened.

I called to him. "Stanley will you come in here a moment?"

"Yes sir," he said.

I said, "Is it true you hit these six boys?"

He looked around the office, "Yes sir," he paused, and said, "but jes' one at a time."

"Why did you hit them?" I asked.

The room erupted as everyone tried to give his own version of the events in the smoke shack.

"I'm asking Stanley," I said raising my voice over the hub-bub. "Well Stanley, why?"

"They been callin' me names ever' since I got here almost three weeks ago an' I took all I could of it! So I asked 'em politely to stop an' they didn't so I hit 'em."

"What kind of names?" I asked. I knew the answer, but I wanted to hear it from Stanley.

"Well, a lotta names, but the one that really makes me mad is 'nigger'," Stanley replied.

"Did you call him a nigger?" I asked, deliberately not specifying who should answer.

Again the room erupted with a chorus of protesting voices defensively accusatory. "Well, he's got no right to hit me. I didn't mean nothin' . . ." Another said, "What I meant to say was Negro." Yet another remarked, "Lucky I decided not to hit him back, I woulda broke his jaw if I had . . ." "He can't stay around here 'cause nobody's safe," and on and on.

"Everyone quiet," I said. "Stanley, are you going to hit these guys any more?"

"No sir, only when they call me nigger. Then I will. I been tellin' 'em I don't want no trouble. But they ain't listened to me, an' I jes' took all of it I could."

"Is anyone going to call Stanley names any more?" I asked.

"Not me man," said Henry laughing. "No sir, I sure ain't goin' to be a fool no more," said another. Repentance was rampant. Everyone assured Stanley of their future good intentions.

We all began to laugh.

"Stanley, there's your answer. If anyone calls you names, you come and tell me . . . right after you hit him. Okay?"

"Yes sir," Stanley replied.

***

Stanley was at camp for about six months. He kept his word and never hit anybody else, and they, I'm sure, never called him names again. He was paroled to his home in Portland. Six months later when I turned on my television there was Stanley boxing in the Golden Gloves Championship. It was the first time I had seen Stanley fight. He was a windmiller who probably could have taken on all six of those boys at the same time.

Stanley continued training at the then renown Knott Street Gym fighting as an amateur. Then he turned pro and won ten straight. He was touted as Portland's next contender, but it was not to be. His career ended with three defeats when he faced more experienced out-of-town fighters. I heard that he longshored for a while, then for a dozen years heard nothing of him.

In the mid-seventies, his son came to camp. He hadn't seen his father in years although he knew that Stanley still lived in Portland. Stanley had hit bottom and disappeared in a haze of alcohol. His son Charles had been raised by Stanley's mother who blamed boxing and big money for Stanley's desertion of his family and disappearance. Charles was heavily into drugs, and ran from camp disappearing into the drug jungle of the street.

The same year I retired, Stanley's grandson was sent to camp. Ron was heavily involved in drugs and gangs. He had no knowledge of his father Charles or grandfather Stanley. I had access to his birth records and face sheets and so had more knowledge of his heritage than he. He showed very little interest in his father, having lived with his maternal grandmother until she could no longer control him after which he was placed in various Children's Services foster and group homes—most of which he ran away from. He was then sent to MacLaren and subsequently to camp. Ron was returned to Mac' for attempting to smuggle contraband drugs, and later that same year was reported in the *Oregonian* newspaper as the shooter in a "gang bang."

In thirty years Stanley's family had disintegrated. Stanley had pride, for a time, but he deserted his family for alcohol. His son remembered him only dimly. Stanley's grandson knew nothing of his grandfather and little of his father and so was destined to destroy himself and those around him as had both his father and grandfather.

Counter and office as seen from the day room;
the dormitory was to the far left.

CHAPTER SEVEN

AN UNAVOIDABLE DELAY

Larry stuck his head in the office door and asked, "Have they called?"

"No," I replied, "No one's called."

"I sure wish they'd get here," he said. "Paul said he and sis would be here by noon. I'm going home with them to stay. I'm going to be paroled to live with them in Albany. I can't wait!"

All the other camp inmates had left on the bus for their monthly visits to their homes and families. Larry was to be picked up by his sister and her husband so I had agreed to wait with him in the empty building for an additional hour.

"I'll call you if I hear anything," I said.

Larry never replied but I heard him as he walked on down the hall from the office.

"Seems like I always sit waiting for someone or something." I thought. "I wonder why they are so late? I didn't know when I was asked to wait that they were going to be so late or I wouldn't have agreed to it."

I took out a magnet and a shiny ball bearing from the miscellany which had been collecting in the top desk drawer as it was abandoned or confiscated from kids. The assortment in the desk was a frequent source of amusement in an idle moment. There were contraband magnifying lenses that could be used to start fires by concentrating the sun's rays, a variety of homemade rings made from flattened pennies and nails and some cheapies ordered from the back pages of comic books. There were small folding pocket knives, and a plastic kaleidoscope and ingenious miniature pistol cigarette lighters. The confiscated items filled the spacious back half of the center drawer. Occasionally an obscene picture, or a deck of nude playing cards, or a tattered *Playboy* magazine would appear briefly. These kinds of contraband didn't stay in the drawer very long. Staff members appropriated such items to save the camp's inmates from such unwholesome influences as sex and nudity.

The desk was covered by a large sheet of glass under which were placed important frequently needed information such as shift schedules, telephone numbers, and emergency procedures. Bored by the inaction necessitated in waiting, I amused myself by attracting the ball bearing to the magnet. Supporting the glass on tape dispensers and staplers enabled me to force the movement of the bearing to the area of the glass that I wanted. Rolling it across the magnets I could watch its course being influenced by the mysterious, invisible force of the magnet.

The telephone interrupted my errant thoughts and dawdlings. "Camp Tillamook, Mr. Graves speaking."

"Mr. Graves, this is Stewart Clarke."

"Yes?" I replied questioningly, hoping that his call would not tie up the line too long since we were waiting to hear from Larry's sister and brother-in-law.

Clarke was the director of psychological services at MacLaren.

"Are you waiting for Larry Smith's family?" he asked.

"Yes," I responded, "and they're three hours late."

"Is Mr. McMillan there?"

"No, he's gone. Only Larry and I are here now."

"Is Larry with you now?"

"He's here, but not in the office."

"Good," Stewart said, "His sister and her husband were killed in an accident near Dolph on Highway 22."

"Oh my God! That's terrible!"

"You tell him," Clarke said, "then bring him to me at the clinic this afternoon."

"I have to tell him?" I asked aghast.

"Yes, it's better for you to tell him in person than for me to tell him over the telephone."

"It'll take me two hours to drive to Woodburn and it's past two now. Will you be there?" I asked.

"Yes I'll be waiting for you."

"Okay," I said, "about 4:30 then," and hung up the phone.

I expected Larry to have heard the phone and to be already at the office door, but he wasn't. I pondered. How do I tell him that his sister and her husband were both killed in an auto accident only two hours earlier?

I didn't know much about his family. I went to the file cabinet, pulled his folder and read. "Larry Smith, age 15, theft, burglary, trespass, assault on a teacher, MIP, harassment. theft 1, truancy, committed 6/19/58 for violation of parole. Father unknown. Mother resides in Salem. Siblings: sister age 23." I continued reading. "Larry's step father is physically and mentally abusive to him. Foster home placements were unsuccessful." I noted he had run away from three. "Recommendation," I read, "that Larry be placed with his sister and her husband who have expressed concern for him. This will provide stability and direction lacking in his mother's home." Not very helpful, I thought, wishing that I hadn't read it.

"Mr. Graves," Larry said from the doorway, "was that call from my sister?"

"No," I replied, "Come in and sit down. I want to talk to you."

"Okay," he said, seeming somewhat surprised at the invitation.

He sat down on the sofa nearest to the desk. I picked up the ball bearing and magnet still lying on the desk and held them apart, allowing them to roll together. Larry looked at me quizzically. Neither of us spoke.

"Your sister and her husband are dead," I blurted. "Killed in an auto accident two hours ago."

He looked at me blankly. "You're a liar. I talked to her on the phone this morning."

"They were killed near Dolph on Highway 22 about noon. It's true. I'm sorry, but it's true."

Larry just looked at me blankly.

What if it weren't true. What if wrong identifications had been made. What if Clarke hadn't confirmed the report any better than I had. What if I were wrong. My mind was racing with disbelief. What must Larry be thinking? I hoped I was wrong, but I hoped they were dead, but if they weren't . . .

"How do you know?" he asked. "Who told you? Maybe it ain't so."

"Mr. Clarke called from MacLaren. The State Police contacted him. I hope it's not so, but we have to go to Woodburn and talk to Mr. Clarke. We need to be sure. Okay?"

I realized that I was rolling the ball bearing across the glass to one hand then to the other. The small whirring sounds seemed to roar in the silence of the office. Larry's head was down so I couldn't see his face.

"Okay," he said dully.

I put the magnet and bearing into the drawer, walked to his side, and gently helped him to his feet with my hand inside his upper arm. "What a shitty way to spend my day off," I thought guiltily. "I had real plans for this weekend."

CHAPTER EIGHT

RUNAWAYS!

A light November rain was falling as my headlights swept the parking lot of the deserted officers' club building across the street from the Camp's barracks. The lights of a car parked there winked out. I had learned to be alert for unusual happenings so I scrutinized the car in passing. It was an older blue sedan, possibly a Buick or Oldsmobile. It was too dark to tell which or whether anyone was in it, but I decided to call the gate security to have them check it as soon as I could get inside to a telephone. I parked and hurried up the steps and through the front door and paused to look through the window in the door at the car across the street. It was too far for me to read the license number. As I headed for the office phone, a hoarse voice, barely a whisper called to me from the top of the stairs. It was Ray the night staff.

I had been out Friday night bar hopping, confident that I could return to my room at the camp unnoticed since it was after midnight. I especially didn't want to be seen by Ray, who was religious "fundamentalist" who was taking correspondence courses for the ministry and who was vocally, vehemently opposed to drinking.

"Hey Jack, come up here will you?" he called to me.

"Okay," I replied hoping that I might evade detection by keeping my distance from him, but since my room was upstairs and I urgently needed to urinate, I couldn't avoid him.

"Tommy is sick," he said. "Can you help me?"

"I'll try," I said, keeping my distance wishing that I had drunk vodka instead of stinking beer. "What can I do?"

"Tell me what's wrong with him and take him to the emergency room if he needs to go."

What a time to come back to the camp tipsy! I held my hand across my mouth and bent over the cot in the small room at the head of the stairs called the medicine room.

Tommy Palchuck lay on his side with his face to the wall. He was an Eskimo youth from the coast of Alaska. He told me that hunters from his village sometimes crossed the ice into Soviet territory in search of seals. Drinking had become a problem to him, and he became a problem to his village so he was sent to MacLaren as a contracted Alaskan youth. Unfortunately Tommy liked being a lower-48'er, and confided to me that he really didn't want to return home. His record at MacLaren included several escapes involving car thefts and drinking. These had extended his six month commitment to well over a year.

"What's up Tommy? Where do you hurt?" I asked. Indian and Eskimo kids don't admit pain or illnesses. They seem to think it unmanly whining so I didn't expect an answer, but hoped for a response such as clutching at his abdomen, groaning, holding a hand to his jaw, or some physical symptom as a clue to his illness.

Response was immediate! His shoulder heaved and he made the compulsive retching sounds of nausea. He was puking, or trying to. Only a bit of frothy slime showed at his lips as his abdomen contracted and his shoulders heaved.

"Can you tell what's wrong?" Ray asked.

"Yeah," I replied. "I need to talk to you out in the hall."

I went around the corner where the white drinking fountain glowed eerily the darkened hall. I didn't want Tommy to hear us.

"He's got the dry heaves." I whispered, covering my own beery breath with my hand.

"Is it serious?" Ray solemnly inquired. "Should he go to the hospital?"

"No," I replied, amazed at his naivete. "He's been drinking. He's sick from," I searched for the appropriate word, "alcohol poisoning . . . well, he's drunk."

The sound of feet stumbling down the stairs interrupted me. Tommy had gotten up and run down the stairs and out the front door.

Running into the medicine room to the window, I saw Tommy get into the old blue car that had been sitting across the street now at the curb in front of the building. As it accelerated, its headlights flashed glistening on the rain spattered blacktop.

I ran to the door and said to Ray, "Have you done a bed check? We're probably missing more kids or have some drunks in bed."

"No," he said.

Only a few feet away was the red fire bell switch which would assemble the camp's population on the sidewalk in front of the building in less than two minutes. I flipped the switch and turned on the hall lights. Kids tumbled out of bed, muttering, mumbling, and swearing. Others shuffled zombie-like. I strode amid the turmoil of the ringing bell and running, cursing inmates to the end of the hall and back checking to see that the rooms had all been vacated.

The inmates lined up by twos on the sidewalk outside the front door as they had been taught making it easier to head-count. Most were fully awake and excited, curiously babbling questions concerning the cause of the commotion. Barefoot and in shorts or pajamas, the rain didn't cause nearly the upset as not knowing why the fire drill had been called. I quieted their questions and counted. Four were missing. I sent Ray to the office for the roster of names to determine which four.

"We're missing Tommy Palchuck, Sonny Tuluma, Eddy Grayeagle, and Eric Wilcoxen," I told Ray. "You put these kids back to bed and call the sheriff's office to report them missing. I'm going to drive around a bit. They can't be too far away."

He led the now wide awake grumbling mob into the building. I had volunteered to conduct my search so that I could avoid close contact which might lead to detection of my inebriation but I didn't envy Ray the task he had ahead in trying to calm everyone down, call in the APB's to the police, and restore order. I rationalized that I really wouldn't be far away, just driving around. Once he had put them back to bed, I could safely return to my own room and bed.

The huge hangars each housed large timber products firms with still others in adjacent buildings. The parking lots held hundreds of autos. This was probably where the boys had stolen the blue car. I cruised up and down the rows of parked vehicles. Suddenly headlights careened toward me, I veered sharply right. We passed with inches to spare.

The main gate of the Industrial Park was guarded to prevent theft of property and wood products. I hoped to get to the gate before our runaways, but they turned the opposite direction so I followed with my lights out trying to keep their tail lights in sight. They headed for the South Gate, unguarded and frequently open in the daytime, but closed and locked at night. Away from the area lights I had to turn on my headlights to stay on the rutted road. Seeing that they were being followed, they sped up and went through the South Gate without slowing. As I passed through the gate I could see that the escapees' car had smashed the chained and padlocked wire and steel double hinged South gate which lay twisted and bent on either side of the now open access road.

I had lost sight of their lights but I guessed they would head for Salem, home to two. I could intercept them by taking the old highway route to its junction with 101 so I pushed my small vehicle to the highest speed conditions would permit. At the junction I cut my lights and waited for them, or better still a police officer to take up the pursuit, to pass by.

Then I saw it, an old blue Pontiac a hundred yards down the highway near the entrance driveway to Simmons' Creek shingle mill. I parked behind it. Seeing no one I walked to the front of the auto. Its windshield and headlights were caved in with shattered glass littering the dash and front seat. Its top had been badly dented.

I returned to my car and got a small flashlight. I knew the anatomy of shingle mills well since I had worked in them from age sixteen to twenty. These mills which cut red cedar roofing are usually built on sloping terrain usually with a dammed creek serving as a mill pond. The log haul and cut-off deck are on the upper level, gravity rollers carry squared cedar blocks to the lower level shingle sawyer who cuts the shingles and tosses them into bins to be packed in bundles at a still lower level. There packers' frames and bins are located at one end of a large storage shed where the shingles are stacked awaiting shipment. This was the most likely place to offer our fugitives concealment.

I quietly followed the graveled roadway, pausing to relieve my bursting bladder at the side of the road. Then I entered the pitch blackness of the storage shed with my flashlight turned off and stood listening. There was a rustle of leaves mixed with the patter of rain on the corrugated metal roof. I backed up toward the side door which I had come in, pressed against the wall, and waited. If they tried to get out that way I could block them. Minutes passed. I waited and listened intently.

Then fear struck. These were four of the largest and oldest youths at camp. In addition to Tommy were Sonny Tuluma, a handsome brown skinned youth of Polynesian extraction, Eric Wilcoxen was an athletic kid from South Salem High School, and Eddy Grayeagle, Warm Springs Indian lad who had stolen and wrecked a tribal council police car. I knew them all well, but they had been drinking. I didn't know how much. Having stolen a car and run away they now were desperate escapees as well. I doubted that I could subdue one at a time, certainly not all four. I stepped outside the door and stood in the rain, thankful that they had not been hiding there.

I switched on my light. As I neared the highway I heard a commotion coming from a small house across the highway. Lights came on. I warily approached. An angry man clad only in loggers' trousers and suspenders clutching a deer rifle was standing in the light from an open door shouting obscenities.

"The bastards just took my pickup," he shouted.

I shouted back, "They're escapees I'm following. Don't shoot. Call the Sheriff's office and report your truck stolen. Tell them the four runaways from Camp Tillamook took it and are heading south to Salem, I'm going to follow them. You got that? Okay?"

He yelled, "Okay." I ran to my car and set off as fast as forty-two horse power would take me.

At Hebo I turned left on Highway 22 still thinking they would head for Salem. I was slowed by the rain slickened pavement. By Valley Junction I was sure that I had lost them and decided to turn around and return to camp. Then I saw police car flashers at that junction.

In a low depression between a house and the end of a bridge was an overturned vehicle lighted by the spotlight on the police car. It was a red pickup truck.

"I've been following them. They're escapees from Camp Tillamook." I told the officer.

"I know," he said. "They took a car from that restaurant across the highway and are gone."

"Heading for Salem I think," I told him. "Two of them are from there, one's from Alaska, and the other from the Warm Springs Reservation. They've been drinking."

"I hope they don't kill anybody," he replied.

"Me too," I said.

When I arrived back at camp Ray was on the phone. "Oh they did. Anybody hurt? . . . Well, that's good."

He hung up the phone, and said, "They wrecked another car at MacMinnville, rolled it. No one is hurt."

"I thought they were heading for Salem," I said.
"They were, but took a wrong turn," he said.
"They sure did," I commented.

***

All four boys were returned to MacLaren School at Woodburn. They said they had been in the shingle mill's storage shed, but left by another door when I entered. Two months later Tommy Palchuck ran away from MacLaren, stole a car, and later died of injuries sustained during a high speed police chase and crash.

A table full of "greasers", October 1963, eating outside after the Columbus Day Storm power outage. The original barracks building in the background was the Officers Club.

CHAPTER NINE

LITTLE JOE

From my second floor classroom windows I saw the work crew crummy at the front door. Such an unscheduled return was not an unusual occurrence and often meant trouble. Once when a work supervisor brought in the crew and didn't get out of the crummy, I became concerned and so hurried down to see if he was ill. After an embarrassed hesitation, he admitted that he wore no trousers. His pants were waving with Old Glory high on a nearby flag pole. Young and strong, he had been wrestling the boys on work crew, one at a time, then two at a time, and finally the "em-bare-assing" discovery that he was no match for a dozen. Luckily one youth could shinney well and he was persuaded to retrieve the work supervisor's trousers from their lofty prominence.

On another occasion two boys on work crew had nailed three of Little Joe's best cigars to the wooden rails of the flat bed truck which necessitated an all day mediation session to soothe Joe's hurt feelings and to enable the two confessed culprits to return to Joe's work crew—if not to his good graces.

I watched, hoping that there was no problem and they would leave soon. Perhaps they had forgotten some tools or rain gear for it was mid-winter and terribly wet with a sou'wester blowing. Little Joe, so called for he was less than five feet tall, climbed out of the World War II weapons carrier vehicle used as a crummy and strode into the front door of the building. Noting with relief that he was wearing trousers and chewing on his usual cigar, I hurried down to discover what had prompted this mid-morning return, for I hoped Little Joe wasn't going to leave the work crew for me to supervise. I had classes to teach.

Little Joe came to work with us after once retiring. He had come to the United States from Switzerland in the late nineteen-twenties as had many, lured by glowing reports of earlier immigrant friends and relatives of the region's rich pasture lands and encircling forested mountains. Joe took a job as a dairy hand. He and his brother were wild youths, but as the two young men matured, they worked hard, prospered, acquired their own herds and land, and, of course, sired large families.

Joe sold his milk to the cooperative creamery association for many years, and so became a well-to-do man with an assured income as a shareholder in the association. He kept a dozen acres to graze half a dozen whiteface beef and sold the rest. He was retired. The United States had been good to Little Joe.

"And why so get up at the Gott-damn crack of dawn to milk de Gott-damn cows and shovel de gott-damn manure all de Gott-damn day?" Joe asked rhetorically in his uniquely repetitious and profane oratorical manner.

(Author's note: To avoid boring the reader, the expletive "Gott-damn" will hereafter be indicated by the (*) mark which may be read as "Gott-damn" or as in the manner of the Watergate Transcripts, "expletive deleted" or, for those readers unconcerned with literary authenticity, simply as "ditto.")

Joe grew bored so he took a maintenance job at the abandoned naval air station and was assigned to manage youths who had been committed to the state reform school for various offenses from truancy to murder. Joe was very defensive about "hees boys", and volatile in both his praise or condemnation of their conduct.

"Dees (*) kids all dey need is a (*) chance. When I was a (*) kid in de old country my professor he say, 'Joe, you never going to amount to a (*)', but you know las' year when I go back to my village I ask where (*) old Professor live . . ."

As Joe rambled on I struggled to follow his narrative through his heavy accent redundantly interspersed with (*)'s. Finding it impossible to imagine that Joe had attended a university, I at first was puzzled by his search for an old "professor" then I recalled that in Europe at that time professor referred to most teachers especially in secondary schools. "So Joe went to Switzerland to visit a former high school teacher," I thought with warm mental images of their reunion, a proud teacher reunited with a long lost student. Since I too was a teacher. Joe's story did flash briefly in my conscience as I regretfully recalled times I had similarly admonished lazy students with such predictions. I vowed that I should hereafter endeavor to be less disparaging.

"So you know what I do?" Joe continued. "I rent de (*) beegest limousine I could find and I drive out to my old professor's house. He live in a shanty with chickens and cows, dirty like a (*) pig on the hill side. He is old and walk with a crutch. I say to him, 'Professor Eberhardt. You say many years ago that I never be success. (*)! Do you see that (*) limousine and driver? In de U.S.A. I have a big farm and herds of fine cattle and automobiles and tractors. I am a (*) rich man!'"

My warm feeling faded as my resolve to utter more words of encouragement grew. Like Scrooge I too could someday face hauntings by ghosts from my pedagogical past.

"What did he say to that, Joe?"

"Nothing," Joe replied. "But anyway (*) I give him a one hundred dollar bill to prove my (*) success."

Joe chewed furiously at his ever present, always unlighted stubby cigar butt and spat specks of tobacco from the side of his mouth. This feat, coupled with his ability to continue a running monologue without missing even one (*) was both fascinating and distracting.

"I'll bet your professor, poor as he was, appreciated the money." I said, conversationally, hoping to put an end to what had been a depressing discourse.

"(*)! My old professor say, 'Thank you very much' but he don't need money anymore. He asks if he can give to a former student who need money to go to university. I say 'sure'. I got so much (*) money I don't give a (*) what he do with it!"

"What a nice thing for your professor to do!" I remarked hoping that Joe could forgive the old man. "Remember the time you brought in the work crew because someone tacked your cigars to the truck bed. You were furious! As I recall you called the boys on your crew 'bastards'. You know, Joe, that's sort of like your professor did."

The point that I was trying to make was lost on Joe for he launched into another angry rampage "(*)! I was mad! You bet I was. Three new se'gars ru'int. Tacked nothing! Dey used (*) 16 penny nails."

"Sometime many years from now, one of those boys may remember you called him a 'bastard'. Maybe he'll rent a limousine and come to see you mad as hell. Could happen you know!"

Joe continued, perhaps noting my warning, "I like dees keeds. You know dat. When dey put my keeds in cells at Woodburn I take dem de cigarettes. Okay? My boys ee's good boys. I like a papa to dem!"

Oh, I remembered all right. The manager of Benson detention unit called to ask me if Joe was okay to admit. I well recalled our telephone conversation, "Hey we got this short guy with an accent here who says he's a staff at your place. Says his name is Joe something. You know him?"

"Yeah," I responded, "He's the work boss, he's okay."

So they let Joe in, and he went up and down the hallway dropping forbidden smokes inside cells on each side. I remembered how the manager of Benson became the butt of everyone's jokes for running a "country club." He in turn blamed me since I had vouched for Joe. It was not a fond memory.

I asked, "What's happening today, Joe? Why did you bring the work crew in early? Got a problem?" I asked suspiciously.

"Jes' came in to get de rest a' my boys, the ones you got in de school, and you too!" he replied, without even uttering one (*).

"Why? You need more help on a big job?" I asked.

"(*)! I come to shake de hand of de next (*) President of the United States of America!"

I knew that 1960 candidate John F. Kennedy was landing for a "whistle stop" at our airport. The receiving line would be long, and the day was soggy. I rejected the idea of taking my school classes over to the airport because of the expected crowds of politically inclined local citizenry. But if Joe was going to take the work crew, I certainly should take the school classes as well.

"Okay, Joe, we will meet you there!"

The airplane landed before we completed the two mile drive to the local airport. I marveled at the size of it. A large crowd of several hundred had gathered. My class members were held well back as a receiving line formed. I recognized several mayors and our County Commissioners. Suits and ties and overcoats were everywhere as a misty drizzle fell. We were held well back by a roped off area which separated the welcoming dignitaries from us onlookers. A ripple went through the crowd as the big plane's door opened just a crack. The air was electric reminding me of the then recent movie "War of the Worlds" when the top of the alien space vehicle slowly unscrewed.

Then the door swung open and there he was—Little Joe, and standing just lower on the steps, the work crew boys, mud covered, wearing their battered rain hats and slickers. They were the first in line to greet the President-to-be atop the stair and at the aircraft's door, Joe, and J.F.K., waved in unison. Joe reached up to put his hand on the President's shoulder, then shook his hand. I couldn't hear what Joe was saying, but I was sure that Professor Eberhardt would be very proud of Joe . . . and that every other word Joe spoke was a (*)!

THE STONER ERA

Serious stoners could be identified by long, often unkept hair, ragged, holey, or crudely patched garments especially old military caps and jackets, jeans, and battered felt hats or knit caps. They prided themselves on being "wasted" mentally and on their ability to "make it" without working. Their communication skills relied heavily upon the words, "cool, heavy," and "man" as in, "Hey Man, makin' by workin's too heavy to be cool!"

Although the stoners grew out of the hippie-flower child era, they lacked their maturity and idealism. An authentic stoner would ingest any chemical substance available to get "high." This was in part due to their addictions, but as often was a part of the stoner's desire for approval by their peers. Such massive long-term addictions led to intense depressions, "flash backs", and self-destructive behaviors.

Most stoners now middle aged, may be found derelicted in every prison and community shelter. Few have been able to adapt to today's "Yuppyism" as have most of the "hippies" and "flower children" of the 60's and 70's era.

CHAPTER TEN

A MATTER OF PRIDE

Armand was a ravenous table beast. He simply didn't know how to eat properly because nobody ever showed him. In all of Armand's life he had never eaten at a table with his family. As he told me later, "I allees jus' catch a bite here and a bite there. Oh, they's food in the fridg'. I mos' allees grab a bowl'a cereal like Wheaties or maybe Crispys an' milk an' sugar an' off to the Center for b-ball or a swim. Ain't never sat down to eat a'fore."

He had become the butt of jokes and crude humor at the table. His way of eating was efficiently disgusting. Armand simply put his face down as close to his plate as possible and shoveled food into it. It was easy to tell where he had sat because his place was littered with debris, bits of food, particles of lettuce, crumbs, or whatever else had been served that meal. Round bare spots on the dark formica table surface showed distinctly where his plate, cup, salad and dessert dishes, and each piece of tableware except for the largest spoon available had been set. Armand only used one spoon and his hands to thrust food into his mouth. He spread peanut butter and jelly on whole slices of bread, folded them in half and crammed his mouth full, chewing and cramming, chewing and cramming. The only mannerful thing about his table manners was that he never talked with his mouth full, he was just too busy eating. Armand could consume a full bowl of soup at a single gulp directly from his soup bowl. Five minutes was enough for him to devour a five course meal including a least one dessert.

Other boys avoided sitting close to him and were quite outspokenly cruel in their comments. Most boys with Armand's physical strength and size would have shut them up with a single growl, but not so Armand. He put up with their comments a lot more graciously than they tolerated his lack of table manners.

At one time the camp had very strict rules about table etiquette which were enforced with often humiliating measures such as banishment from the table to the floor or outside. Table service was replaced by a wooden trough. But times had changed; ridicule gave way to indifference. Our new administrator was an avowed Berkley liberal who directed that students' social skills or language should not be corrected which he interpreted as violations of personal liberty and freedom of speech. I disagreed, but still I was glad no longer to have to nag, "Eat your vegies", "Sit up straight", "Don't talk with your mouth full." I much preferred relaxed dining as far away from the "animals" as possible rationalizing that kids like Armand would probably never eat another "sit-down meal" after they left camp and manners don't matter at Burger King.

They mattered to Mr. Johnson. Howard Johnson was our other teacher at the time. He was a crusty old gentleman who had come to us after retiring from Benson High in Portland and before that he taught at Cass Technical High School in Detroit. Students there were later to become some of the biggest names in the American auto industry, and in 1920 Howard had been posted at West Point during the period when Douglas MacArthur was Superintendent there, not as a cadet, but as a sousaphone player in the band. He had traveled with Henry Ford and Edison on their famed camping trips and had been a member of the University of Chicago math team who had developed the atomic bomb. Mr. Johnson, teacher—companion to some of the great men in history, had determined that

Armand Lewis was going to learn good table manners. Who dared argue with him?

When I heard Mr. Johnson's first outburst I was sure we were in for trouble. His tirade declared, "Armand, if you're going to slobber all over and slop food on me go sit at another table. You eat like a pig, don't ever sit by me again!"

I looked across the room, somewhat amazed at Mr. Johnson's uncharacteristic outburst. He was seated at one end of a table with Armand to his left. I half rose to assist Mr. Johnson for I was sure that Armand wouldn't be able to keep his temper, and I feared he would retaliate with his fists.

Armand had always been courteous and pleasant to me. Even the time I caught him smoking in an upstairs closet and put him on report so endangering his home visit privilege. He had not been angry or even sullen. He readily admitted blame. Johnson's verbal assault would probably have brought an angry response from most of our students, but Armand was different.

He was sixteen and an All-State footballer in Portland's Interscholastic League. His coach wrote him often and visited him twice, "Like my very own son. . . .," he said, but I'm sure he omitted, "and he can run the hundred in under ten seconds carrying not only a football but two or three would-be tacklers."

There was a stunned long silence. Everyone sat expectantly awaiting the storm. Then amazingly Armand answered in a low and controlled voice, "Yes'sur Mr. Johnson."

"Eat solid food with your fork," Johnson said. "Hold it like this, not like this." Mr. Johnson showed him how to hold his fork.

Armand picked up his fork and clumsily holding as instructed speared a mass of macaroni and cheese.

"No, no, Armand. Place your fork under the food like this and separate it when it globs together. Like this," Johnson said.

Armand did as instructed. The macaroni fell off his fork. The other boys at the table snickered.

Armand said, "Mr. Johnson, Sir, Dis this is hard doin'."

"Just takes practice," Johnson replied. "Now I am going to show you how to use your knife and fork to cut food."

After lunch I jokingly asked Howard if he wanted me to assign Armand to sit by him each meal. He responded emphatically in the negative making it quite clear that Armand and he would both be happier apart at mealtimes.

I was sure that from that meal on Armand would sit as far away from Mr. Johnson as possible, but he wasn't. He always sat next to Mr. Johnson who was harsh in his criticism, but patient and thorough in his instruction. I often thought as Mr. Johnson berated him, "Either the kid's starved for attention or he's some sort of masochist. Why does he put up with Johnson's verbal harassment meal after meal?" Armand continued to sit near Mr. Johnson; as his table manners improved Johnson's tirades slackened.

Then Armand came to me and said, "Mr. Graves, you know's Missta Johnson is helpin' me wi' m' manners. Will you help me learns ta talk betta?"

I was astonished! How could I as his English teacher refuse such a wonderful request? I replied, "Sure I will, Armand."

I had long before given up correcting students' "aints" and "we was'es" for two reasons: It became a monotonous attention seeker's word game, and secondly, the process of repeated correction inefficiently used time needed to communicate other kinds of information. I believed, and still do, that once a teacher instructs a student, it then is for that person to decide whether to make the conscious effort oral improvement requires.

But here was Armand asking me to, so for the remainder of his time at camp every chance I got I said, "Not I ain't, Armand, I am not", to which he would ask about "I isn't" which is difficult to explain the "why nots" of, but Armand was very patient with me and I with him. I devised special practices in usage in which Armand and I recited over and over, "I am, you are, we are, he-she-and-it is . . ." We did this about thirty minutes each afternoon when school was over for the day. I grew bored with it and thought that Armand would as well, but he persisted and became quite a polished gentleman in both his manners and speech.

***

Armand stayed at Camp only seven months, the off-season, as agreed by the court. When he left camp. Armand was very different, at least so far as his table manners and language correctness were concerned. He went on to the University and the pros where I followed his gridiron exploits. In subsequent years he has been a successful television sports commentator.

Carey J. Huckaba, work crew supervision for state parks seated, Floyd Case leaning over desk.

CHAPTER ELEVEN

a.k.a. ALEC.

(a.k.a. is a rap sheet notation meaning "also known as")

Bob Patton arrogantly announced that he had no need for education since he intended to become a professional thief. We were studying income tax returns that day so I pointed out that if so he had better learn to file his tax forms for some infamous thieves were sent to jail for income tax evasion rather than their other crimes because they had failed to file tax returns.

"My old man is a successful thief, and he's never made out a tax form in his life."

"How much has he stolen?" I asked.

"He said over a hundred thousand dollars last time I visited him," Bob replied.

"That's a lot of money! Did he embezzle from a bank?"

"No, he started as a burglar but then being smart enough to go for the big money began fencing hot stuff for other guys."

"You said last time you visited him. Are he and your mom separated or divorced?" I asked, picking up on a previous statement Bob had made.

"Yes, my mom ran off with some guy while my old man was in the pen. My dad says she was a bitch in heat anyway. Always fooling around. I live with my grandmother in Albany."

"So your dad's been in the pen," I responded, uncertainly.

"He's there now. That's why I live with my grandmother. He may never get out."

"Why," I asked, "did he kill somebody?"

"No," Bob replied. "He's a habitual criminal. They can do that to you if you're convicted more than three times."

"How long's he been in prison?"

"Three and a half years this time, but altogether over twenty years," Bob answered.

"Well, it takes me about eight years to earn a hundred thousand . . . make that ten years, after taxes," I said remembering where our conversation had begun.

"What did he do with all of his money?" I asked.

"Women and booze! He had a really great time last time he was out. But he's broke now. The lawyers said they would get him off. Instead they got all his money. When I go to see him, I take a couple of bucks so he can buy cigs."

"I figure he's been paid about sixty cents an hour," I said. Bob looked at me with narrowed eyes, as I continued, "and even then he's spent most of that on lawyers, right?"

After a long pause Bob said, "How do you figure sixty cents?"

"Twenty years in jail, twenty-four hours a day, seven days a week, thirty days a month. I didn't figure leave time or time off for good behavior. Figure it out for yourself. Here's a pencil."

"Well anyway he's had a lot of fun!"

"What do you think he's doing tonight?" I asked.

"I don't want to talk about it anymore. Give me the pencil."

"Sure." I handed him the pencil.

***

Bob was a capable student. He was very thorough, almost to being a perfectionist. He thrived on praise so I kept praising and he kept working. One day he brought me a poem.

"This is good. I like it, but why is it signed Alec.?"

"That's my other self."

"You've put a period after Alec. Is it an abbreviation for something or is it a pen name?" I said, "Lots of writers use pen names."

"Alec. is my other self, I'm two people."

"Is Alec. your middle name?"

"No, it's just a name that I like the sound of. I make up stories about Alec. He makes me feel good."

"Okay, if you want, you can be Alec. when you write, but remember it's against camp rules to use an alias instead of your real name in outgoing or incoming mail . . . also to avoid confusion, I'd prefer to call you Bob. Okay?"

Because of his avowed interest in writing, I designed course called "Literature and Creative Writing" in which various classic works and their authors were studied as were literary terms and devices. The last assignment was a creative writing project of the student's choice.

"These poems are excellent, Bob. Let's send some in for publication," I said.

"Where would we send them?" he said.

"Well some to *Northwest Magazine* section of the Sunday *Oregonian*, and the *Nehalem-Times* for starters. We can compile them and self-publish here for your friends using the Ditto. Okay?"

"Sure," Bob said.

We sent several poems to different publications, all were published. A nice little verse called "The Surf Exults" was published in the *Nehalem-Times* as a full page feature with a photography essay surrounding it:

Foamy furrows form in the sand,
Where some see only a gritty band.
I like to think thoughts of times,
Long past in these changing lines.
The shape of every minute grain,
Tells tales of stony sandy strain,

Tossed, dashed to bits around about,
As the surf exults its salty shout.
The mental pictures it portrays,
Speak to me of ancient former days,
Of eons endlessly one by one gone,
Bit by bit breaking its stony bond.
The ocean's vast watery expanse,
Sings songs of endless romance,
That you and I may someday find,
In romantic reaches of our mind.

Alec.

For his writing project Bob wrote a novel. He wouldn't let me see it until he had completed it. He had written it entirely in ball point pen on lined filler paper, over 200 pages.

Bob's story was entitled *On The Run*. It told of two runaway youths from a camp very much like ours. They were involved in various adventures which were potentially hazardous to them and others. One youth was named Alec and the other Robert. Alec's coolness and leadership kept the pair from several disasters until the last chapters when Robert impulsively murdered a motorist who had given them a ride. They were caught and stood trial separately for murder. Robert was acquitted, but Alec was found guilty and sentenced to life imprisonment.

The situations and dialogues of the book were trite. The semblance of characterization and imagery was apparent. I proofread and Bob edited his book earning a semester's credit in typing. Bob was released while preparing a final draft of *On The Run* which he wanted to submit to a publisher.

He went to a private care facility in Portland. This corporation operated several juvenile resident halfway houses in the metropolitan area. I told Bob before he left that when I was in Portland, I would pick him up to have dinner with me and my family.

About a month after Bob left camp, I called and made arrangements to pick him up on a Friday evening. When I arrived at the house I found boys engaged in washing windows and cleaning. One youth was sweeping the floor with a toy broom about a foot long, somebody's idea of a joke I supposed. Bob was nowhere to be seen.

When an adult appeared, I said, "I'm Jack Graves. I called day before yesterday about taking Bob Patton out to dinner tonight. Is he around?"

He answered, "Alec's confined to his room until eight and has lost his privs. He can't go with you."

I was suspicious of this individual especially since "Youth Programs" was noted for hiring unqualified people whom they could pay far less. This young man was probably in his late twenties. He had long hair tied with rubber bands at the back of his head and was wearing blue jeans and a very dirty sleeveless t-shirt. An upper arm was heavily tattooed and he wore a small gold stud in his nostril.

"I'd like to speak with him," I said, pulling out my wallet and showing him my CSD identification card. "I'm a teacher at Camp Tillamook. He was in my class there."

The man hesitated and then said, "Okay, I'll get him."

He went down a narrow hall beside a curved banister of a staircase and soon returned followed by Bob.

"Hi Bob," I said, "I hear you're confined to your room."

"Here I'm Alec.," Bob replied, then he turned to the man I had met and said, "Can Mr. Graves and I talk on the porch. You can see us through the window," motioning to a large low window which commanded a view of the porch through torn lace curtains.

When we got outside Bob said, "Yeah I'm confined, but it's a bad scene. I'm accused of taking another guy's 'stash.' Believe it or not, I'm the only straight kid in this dump! The biggest pusher in town is Bill, the jerk you just met."

"I thought he was a staff," I said.

"He is, and pushing hash, LSD, and pot. I've got to get out of here. Can you help me?"

"You can help yourself. Turn 'em in. You're the one who must do it," I told him.

"They'll kill me."

"Tell your parole officer."

"What'll he do about it?"

"First thing he'll do is get you out of here. Then there'll be an investigation."

"I'll have to think about it," Bob said.

"If you want, I'll call your P.O. Monday, and tell him you want to talk to him."

"Okay," Bob said, "but don't tell him why."

"Alec, you'll have to come in now. Ten minutes are up," said the staff, his nose stud twinkling as he spoke.

I said to him, "I'll come back at eight so Bob . . . ah, Alec, can still come out with me to dinner. Is that all right?"

"No," the man said, "He can't go out with anyone."

"What'll you do after eight?" I asked Bob.

"Hit the street," he replied.

"Can he do that, and yet not be able to come with me and my family for dinner?" I asked the young man Bob had called Bill.

"Yeah, he can go to the mall."

"That doesn't make any sense to me," I commented.

***

Monday I called Bob's parole officer. Wednesday Bob came back to camp.

"You told your P.O.?" I asked him.

"Yeah, it's just like you said. Bill is being investigated. I'm here for two weeks, then I get to go to my grandmother's house in Albany."

***

The day after Bob left, the mess at the Fourteenth Street house broke. Bill was arrested as were a number of inmates who had been running drugs in the schools for him.

Bob went to his grandmother's home. A month later when I asked his parole officer about him, the P.O. said, "Oh, I thought you'd heard. He beat the hell out of his grandmother and he's in jail. I'll probably remand him into the adult corrections system."

"Do you know he's also known as Alec?" I asked.

The parole officer replied, "Yes I know that a.k.a. Did you know that alias is his father's name?"

Day room as seen from between the classroom and the office.

CHAPTER TWELVE

PURGATORY

It was the "Age of Aquarius", of L.S.D., flower children, and stoners. The Promised Land was Berkley. They traveled in vans. VW's and old yellow buses with torn curtains at the windows. Some hitched confident that they wouldn't wait long before a brightly painted, flower emblazoned bus would stop, eager to help a fellow pilgrim, and with an exchange of peace signs welcome the weary fellow traveler with, "Hey, jump in man, wanta hit?"

But for some purgatory intervened. . . .

***

"Hello Jack, you wanta new student. We got one to send you."

"A dump job! Right? This kid eats teachers," I laughed, not from amusement so much as sarcasm. It was common in our system to send incorrigibles from one program to another. An incorrigible in one program will likely be as difficult in another but still the tactic had succeeded often enough with some youths to have a degree of validity. I had been "dumped on" many times starting a dozen years before when Gary came to camp after having climbed the MacLaren water tower and pissed on the Director of Home Life who was pursuing him up the ladder. Often there was nothing that I could do, but grumble and devise coping strategies to deal with yet another obnoxious student for as long as he lasted at Camp. Still I wanted it known that I knew why such transfers were made.

"No, we don't have the facilities here," Dick Peterson, then Camp Florence's Counsellor and a year before ours, responded.

"Our cages are all full too," I interrupted.

He didn't laugh, "Books I mean, you know Emil's geared up to teach remedials, and this kid's a genius!"

"Oh come on," I said, "The last bona fide genius I had is doing thirty years in the state pen for manslaughter. I don't care how smart he is, I want to know if he's gentle—like does he throw spit wads . . . or chairs?"

"I swear he's a sweetheart. He just needs to be in your class 'cause you're equipped to teach him better than we are."

"How many credits does he have?" I inquired.

"Don't know, he was just traveling through from Seattle to Berkley when he got arrested," Dick answered. "A Coos County judge said, 'Ah'm gettin' tired of these city kids coming down here and smokin' pot with our locals who then gotta' face the music while city-boy leaves with no consequences so ah'm goin' to make an example of this one.' But," Dick continued, "since Peter had no 'priors' he sent him here as a direct commitment."

Direct commitments were being used by judges who wanted lesser offenders kept from contact with the hard core delinquents at MacLaren. While the idea was commendable it frequently failed because these non-institutionalized boys were harder to handle than the hard cores were and frequently ran away or had to be placed in the stricter detention available at the main institution at Woodburn. Now I was certain this kid was a dump job.

"So what makes you think he's a genius?" I asked.

"The newspaper. Didn't you read about our whale a couple of weeks ago?"

"Sure, the one that came up on South Beach and they blew it to smithereens, and the larger smithereens crushed some cars parked half a mile away."

"That's the one. Well when it came up on the beach, OSU sent several oceanographers who were standing about near it arguing whether it was a blue whale or a gray whale just as several of our work crew kids and some news reporters came close enough to hear. Peter correctly identified it as a gray whale."

"Sounds like simple cases of color blindness to me, Oregon State University should have sent the art department instead of oceanographers," I asserted, chuckling at my wit.

"Well," he continued unamused, "a big argument resulted, and Peter isn't one to be contradicted, and . . ."

"So this *is* a dump job then!"

"But Peter was right, and when the news people discovered he was right and the professors wrong, they made a big thing of it so now the kid thinks he's too smart to be in our school because he's an oceanographer, an' all we teach is readin' an' 'ritin'."

"So what makes you think I'm an oceanography teacher?" I asked.

"Make something up. You're supposed to have the school program geared to advanced students," he replied.

"Yeah ten years ago, now I just get dumped on. But okay I'll take him," I conceded, adding, "but remember you owe me one!"

"Hey that's great Jack! I'm sure you'll do wonders with him."

***

The next morning there was Peter. I was surprised to discover that he had been delivered to us so soon. His early arrival made me even more certain that Camp Florence wanted to be rid of him as soon as possible, but the blocky brown haired youth looked like any other kid to me, and since our count was then the lowest in the Camp's history, I welcomed any student, especially a genius!

"Hi, I'm Mr. Graves," I introduced myself, "and you're Peter. I heard about the whale. How'd you know what kind it was?"

"Read about 'em and saw pictures. I'm interested in oceanography. That's why I'm here . . . to study oceanography," Peter answered self-assuredly.

"I'm sorry to say that I don't know much about oceanography, but we can learn together. I have some books we can use and when necessary we can go to the county library or order books from the state library if we need to. But now what I'd like you to do is plan your course in oceanography. Take this paper and pencil and," I pointed to a back table in the room, "write down what you want to learn about the ocean."

"Okay," he said, and sat down at the table.

In five minutes he was back with one side of his page filled with ideas.

"First I see you want to go to the beach to collect specimens for study."

"Yeah, that's a good place to start."

"I think you're right, so after lunch we'll do that," agreed. "I know a place where a road embankment at Bay Ocean has exposed lots of old shells. Some fossilized."

The beach was windy and cold. I parked and Peter and I walked along the roadway. I was looking for a good place to dig, Peter was enjoying the stroll. He lit a cigarette.

"There's a good spot," I said, pointing at many small white flecks in the brown sandy earth and handing him the shovel.

"Naw," he said, "this isn't the place. Let's look for shells on the beach. We might even find another whale."

We walked along the beach stopping twice to examine possible finds. Finally Peter sat down on a log and cupping his hands against the wind, lit another cigarette.

"It's time for you to decide whether you want to collect specimens or smoke," I said. "If you want to smoke, you can do that back at camp, right?"

"Yeah, why not? This is boring anyway," he replied sourly.

Peter's oceanography soon ended. When I tried to get him to read from the few marine science books we had available. He scanned the pictures, but refused to read the books or write the required reports so in the end I put him back into the three r's—assigning him world history and grammar classes as well as math and science classes in then teacher Alex Berlin's classroom. Peter was easily bored with basics such as composition and history, but did reasonably well only because I repeatedly assured him achievement was required if he were ever to regain his lost freedom.

At that time the whole system was in turmoil. "The Gault vs. Arizona Decision" had been interpreted as requiring that juveniles be accorded the same legal protections as adults. Subsequent decisions and the Oregon Attorney General's opinions further reasoned that equal protections must result in equal penalties. These rulings were to have far-reaching consequences, for Oregon had no direction. MacLaren's infrastructure had self-destructed. The Superintendent had directed that inmate students were entitled Constitutionally to "free speech" including especially obscenity and profanity. Further he directed that: "Adult staff members should be sufficiently mature to accept verbal abuse from students without resorting to angry retorts or other retaliation." Our Director was ill most of the time, and his assistant had just read *I'm OK, You're OK* which he correctly interpreted as "making students feel good about themselves." Unfortunately neither his nor my students' idea of well-being included class work. On one occasion when I kept two younger students in during a recess period, he called me into the office and told me that his first official act when he became Camp Director would be to fire me! I told him that should he ever become Camp Director I would quit anyway.

Ironically a few months later when we were the only two staff in the building, I received a call from the Assistant Superintendent ordering me to place him on immediate suspension so as it happened, I fired him!

I remember during this period one of the worst days in my thirty-three years as a teacher occurred. I was alone in the building with a dozen or so students when one ran away. He was seen running across a field and two of the larger boys volunteered to catch him so I sent them after him. This was the first and last time I ever used "hiker-chasers". Outside I watched them pursue and capture him, surrounded by foul-mouthed inmates calling for his captors to beat him while obscene "free-speech" Woodstock music lyrics blared from an outdoor loudspeaker.

When the three returned to where I stood, the hiker had been kicked and punched repeatedly as he struggled to be free. I ordered that he be released and told him if he still wanted to run away he could leave without pursuit because I would like to go with him. That humor broke the tension of the moment, but my offer was not entirely in jest. I ordered him freed and offered to allow him to leave without pursuit. Fortunately he stayed to become a successful student both at camp and according to his parole counsellor Bob Warner later at The Dalles High School.

One day Peter was looking through the book shelves and came across a small book I had picked up at Powell's on computers which was titled something like, *Talking to your Computer*. He was excited and subsequently spent so much of his class time scribbling cryptics. I only allowed him to "program" after he had completed his other class work which proved an incentive "carrot" to get him to do his class work. A few times I looked over his shoulder curious about what he was doing. He knew about computers for he had started using them in his home high school the year before, but his best explanations were lost on me since I had never even seen a computer and had only a foggy notion of what use they were.

One day, as happened frequently, there was a mass "sick-out", a commonly used subterfuge to avoid going to work. Since it was impossible to determine who was really sick and who was not we had adopted a general rule that when a boy stayed off work crew he had to spend the next twenty-four hours in bed. These "twenty-fours" were miserable days for the faking youths who soon grew bored and restless, but even worse for the staff who—usually me—had to try to keep them in bed, maintain order in the dorm, and, of course, hold class also. After the first two hours in bed it was easy to tell who was really ill for they slept quietly—not so the fakers.

Peter was one of half a dozen who had taken "24's." I was trying to teach thirty feet away in the classroom. I repeatedly shouted for the "sickys", as we called everyone in bed, giving even the malingerers the benefit of the doubt, to "Quiet down." Several times I went into the dorm to determine who was the source of the disturbances. This soon became a game with all kinds of ingenious deceptions devised, strange sounds, small articles and pillows thrown, etc. I accused Peter of being the source of a particular disturbance. Rightfully or wrongfully didn't matter for he was openly snickering causing me to lose my temper, I grasped him by his shoulder propelling him toward the office causing him to become indignantly insolent. In the office I isolated him from the others and gave him poor grades restricting him from town visit.

After that we weren't chummy, but most mornings we would chat about events reported on the sports pages of the *Oregonian*. Peter was a great fan of the Blazers, especially Geoff Petrie. Each morning he and another youth named Fred would vie for the sports section so they could read the "Blazer Scoreboard" and critique the game of the night before. Camp Director Dick White and cook (later teacher) Doug Hogland and I often entered into these hypothetical "bull sessions" about which Blazer players should be replaced by whom. It was great conversation since we had the whole NBA to pick from as we verbally built a super Blazer team.

My longest conversation with Peter was the day he left. We were standing together at the office counter doing his clothing and property check-out. I asked what he intended doing when he got home. For me this was usual small talk to occupy idle moments with students waiting to go home. Peter replied that he hoped to write software programs with an old high school friend. I was skeptical having heard many grandiose plans from previous students, only to find these well-intentioned youths back in my classroom in a month or two. "Too bad," I thought, "that he isn't able to read blue prints or weld so he can get a good job in Seattle. I doubt that there's much demand up there for a computer 'nerd'."

Me—Jack Graves—seated at my desk.

CHAPTER THIRTEEN

I'M WITH MY BROTHER

It was Wednesday morning. I pressed the buzzer to start class as I did every morning then went in and sat down behind my desk. As usual students were already seated and working. Doug Hogland, now a teacher, and I were busy just trying to keep up with a full house of thirty busy students.

Dick Peterson had become superintendent and he made it clear that education was among his top priorities for institutionalized youth. His policies were those of behavior modification and reality therapy which simply meant that youths were rewarded for good performance in a very structured setting. Goals to be met prior to release were clearly established for each student in conferences including him, the teachers, the camp counsellor, and his parole officer, with input from his parents upon his arrival. Misconduct was clearly defined and consequences for defiance exacted were swift and certain. Every camp student knew that should he "screw up big-time" such as "bringing drugs into camp" or run away, he would be returned to MacLaren's Woodburn facility, "campus" as it was euphemistically called, for "sixty days hard-time"—no home visits or privileges. Then he would be required to return to camp to start his program again with no release possible until he succeeded in meeting all his goals for release.

Three new boys were in the classroom with six familiar faces already seated. Two of the newcomers were seated and a third stood just inside the door.

"Take a seat anywhere you want. I'll get around to talking to you as soon as I can," I said to the one near the door. "Meanwhile you can read something, a magazine or newspaper. Okay?"

"I'll sit here," the tall slender youth said, taking the chair next to my desk, "I'm going to be teacher's pet."

The lad was sitting directly in front of me. Not knowing his name, I said to him, "I'm Mr. Graves. What's your name?"

"Joe," he replied.

"Joe what?" I asked, not surprised as I had done this same thing this same way hundreds of times.

"Johnson," he mumbled.

"Where you from, Joe?"

"Kaiser," he said. "It's near Salem."

"Yes I know," I replied. "My brother-in-law teaches there."

"Oh yeah? What's his name?" Joe asked.

"Harry Pease," I replied, "Do you know him? He teaches at Cummings School."

"No," Joe said, "and that's good 'cause if he's a teacher in any of the schools I went to he probably kicked me out of school at least once, maybe two or three times."

"What class are you in?" I asked, ignoring his comment.

"What do you mean class?"

"I mean sophomore, junior, like 10th or 11th."

"I dunno. I got two credits since I been in school."

"How old are you?" I inquired.

"Seventeen."

"Well, it's time to get some credits. Right?"

"I guess so," he mumbled.

"What do you want to start with? Math, science, English, or geography?"

"I don't care," he replied.

"That's just the answer I was wanting," I replied, rising from my chair. "Let's go to the classroom next door so I can introduce you to Mr. Hogland, the world's greatest math teacher!"

The next boy I talked to was very much like Joe, only he wanted to start with world history. So I got him a book, notebook, pencil, and showed him how to get started. This had taken twenty minutes. The tall kid I had asked to sit down was reading the sports page.

Well I thought, at least he can read, and said to him, "I'd like to talk to you now."

"What's your name?" I asked, standing to shake his hand, and to direct him to the chair.

"Jerry Harrison," he replied.

"Where you from?" I asked once again.

"Portland," he said.

"What class are you in?" I inquired as I scanned his transcript.

"Junior," he said, "but I dropped out. I don't go to school anymore."

"You'll have to be in school here," I said, not surprised since many of the youths I talked to had dropped out.

"What's your commitment offense?" I asked him.

"Truancy," he responded.

"You're just here for truancy?"

"Yeah, the vice-principal called and threatened my dad so went to school the next day just to cuss him out, and I did. I guess it made him mad. So here I am."

"I see from your transcript that you have fourteen credits, and your grades aren't bad," I said. "Why don't you plan to graduate this year. You could probably earn six credits here."

"School bores me," Jerry said. "Teachers just blab and blab, and don't say nothin'. That's why I quit. I didn't like what they was trying to teach me."

"So what were they trying to teach you?"

"I don't know, I didn't listen to 'em."

"Do you want to graduate?" I asked him.

"Yeah," he said, "if I can study what I want to."

I replied, “I’ll show you some interesting credit courses, but remember I’m not going to teach you.” He looked at me quizzically. “I’m a learning facilitator, and you’re a learner. I just show you what’s required to graduate. The rest is up to you.”

Jerry stayed at camp for six months the first time. He worked diligently and became a model of success. He was paroled to his father and was within only two credits of graduation when he returned to attend a Portland high school that spring.

***

The following September I was talking to Virgil Thompson who was Jerry’s parole officer, “You want Jerry Harrison back?” Virgil asked, “he wants to come back.”

“Well I’m sorry to hear that he’s in trouble,” I replied.

“He’s not, he wants to come back here to go to school.”

“He’s coming back for truancy again then?”

“No he’s eighteen now, and doesn’t have to go to school, but he wants to graduate. He wants to come back voluntarily.”

“Can he do that?” I asked.

“Only if I revoke his parole, and I can do that until he is twenty-one years old,” Virgil said.

“Well, if he wants to come back, it’s okay by me,” I said. “He can graduate by Thanksgiving if he works as hard as he did.”

Two weeks later, Wednesday morning Jerry was back.

“Ready to go to work?” I said.

“Yeah, I want to be in the Marines by Christmas.”

Jerry graduated on his schedule. The day he left camp he came to me and thanked me for the help I had given him. As we chatted, he said, “My brother wants to come to school here next semester.”

I laughed and said, “This is the state training school for delinquents. He can’t just enroll here whenever he wants. He has to break the law . . .”

"He's not going to school so all he has to do is go down to Madison and cuss out Shaeffer, the Vice Principal . . . just like I did. Shaeffer will see that he gets here."

I laughed and said, "Tell him I'll be here waiting for him."

"I'll tell him. His name is Frank."

***

Another Wednesday a couple months later, two new boys were sitting in my classroom when I entered.

One of them, a tall young man with broad shoulders got up, offered his hand, and said, "Hello Mr. Graves, I'm Frank Harrison. I have twelve credits, and I've come here to graduate so I can join the Marines like my brother Jerry."

I said, "You're joking aren't you. Virgil put you up to doing this didn't he?"

"Virgil is my parole officer." Frank said. "He did recommend that I come to camp when I asked to because Jerry did so well here in school."

"Well how did you get sent to MacLaren? What did you do?"

"Quit school and cussed out the Vice Principal."

"Sit down, Frank, and I'll show you how to graduate. Do you have any brothers at home?" I said smiling.

"No, why?" he responded.

"Well good, when you go home, promise you won't tell any of your neighbor kids about this diploma mill we're running. Okay?"

Frank, who proved to be even a harder worker than his brother Jerry, had been at camp seven months and was within a credit and a half of graduation when disaster struck!

***

"Frank, you have to leave Camp and finish school at home," Virgil Thompson told us after calling us out of the classroom.

"But another month and I'll be finished. Can't I stay that long?"

"No, you have to leave by Friday."

"Who says so?" asked Frank.

"The Supreme Court of the United States says. In a decision last year called Gault vs. Arizona the court ruled that juveniles cannot be treated differently from adults in sentencing," Virgil explained. "You were committed for truancy. Since an adult could not be sent to jail for truancy, you can't be held in MacLaren. We are releasing fifty-five boys by noon Friday, and you're going to be one of them."

Frank argued, "But I don't want to go. I've only got a credit and a half to complete for graduation. Isn't there any way that I can stay another month?"

"Not unless you have committed a felony crime you didn't tell us about."

"Like what?" asked Frank.

"Robbery, burglary, car theft . . ."

"I took my dad's car once without asking. Boy was he mad!"

"That doesn't count because he didn't press charges."

Frank left camp on Friday as scheduled. He said he was going to catch the next bus back. I told him if he did I would have to call the police and have him arrested for trespass. That sounded pretty good to Frank until we found out that trespass was not a felony offense and wouldn't cause him to stay at camp so he reluctantly left for home that Friday.

***

On Wednesday morning, a month later. I looked over our new arrivals as they entered the classroom, and there was Frank.

"You're back," I said. "I hope you didn't go out and steal a car or something . . ."

"Well, sort of, I talked my dad into pressing charges for the time I took his car without permission. He and Virgil went to court with me so I could plead guilty to auto theft."

"You'll never get into the Marine Corp. with a record like that," I said. "Did you think about that?"

"I didn't," Frank said, "but the judge did. The day Virgil sends him proof that I've graduated, he tears up my record. It will never be recorded."

"Welcome back, Frank. Let's get to work! Okay?"

"Okay," he said.

***

Frank and Jerry both were successful Marines earning honorable discharges after five year hitches. Jerry came back to Portland and last I heard was an electric lineman. Frank stopped by one afternoon seven years after he had left camp. He was accompanied by an attractive young woman who was his fiance. Both were students at the University of Oregon.

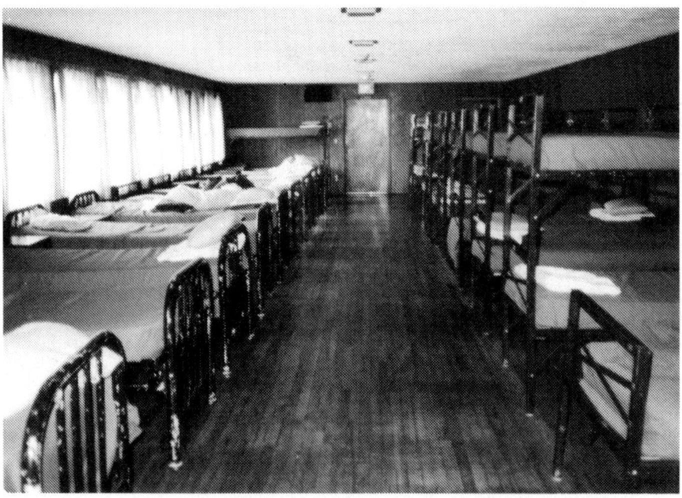

Looking into the dormitory, capacity 25-30.
Note the double bunks on the windowless wall.

CHAPTER FOURTEEN

HARD LESSONS

Curtis Taylor and Anthony Carver were ghetto youths who had learned the aggressive hostile avoidance behaviors common to many, if not most delinquent, disadvantaged inner city males. In angry verbalizations they use aggressiveness defensively. Such youths are either unwilling or unable (many find it expedient to portray themselves as unable) to compete for attainment of societal and school goals and so avoid competition through hostility and belligerence toward authority figures, especially teachers. Teachers are often intimidated, but even when they are not, they find it impossible to devote such time and attention as necessary to help these students solve their individual dilemmas so these youths are permitted, even encouraged, to vegetate in junior high and by their second year in high school to drop out.

Trying to help such kids is one of the least satisfying and most difficult aspects of correctional education since they are not appreciative at the time of intervention. The public sector education deals with youths like Curtis and Anthony by rejection (expulsion further diminishes self-esteem) or by placing them in alternative vegetative programs in remote classrooms in which they only fall further and further behind). Many correctional teachers cope by enforcing rigid disciplinary measures and using rote learning methods such as workbooks which may sometimes force students' grudging class work, but not long enough to establish a level of confidence and competence sufficient for them to acquire socially acceptable coping mechanisms and competitive behaviors.

When society and the courts recognized minority rights, correctional facilities found it impossible to provide programs suitable for such youths thus isolating them and at the same time insulating administrators and teachers from the many frustrating legal and social complications necessary to educate and rehabilitate them. Such were the beginnings of today's punitive "lock everybody up" correctional policies which are now supposedly justifiable in terms of societal protection.

***

Curtis' first day in class he was totally uncooperative. He refused to talk about class work or to take placement tests. After an angry exchange, we settled into a waiting game. Curtis sat and glared an unblinking, surly stare. I informed him that unless he at least made a reasonable effort, I would recommend that he be given no privileges that week and the week after if he showed no progress he would be returned to MacLaren. The next several days he was so disruptive that I had to insist that he leave my classroom, and provided him with a seat outside the classroom in a location where I could observe him. His hostility was evident in every contact we had and even when there was none he lurked nearby glaring and making physically hostile gestures.

I had parked my pick-up just outside between the storage shed and the camp building. A convenient spot but well screened by the building until the last few feet. That Friday, as I rounded the building there sat Curtis Taylor on the hood of my truck, his feet dangling, glaring at me in a manner now all too familiar. I knew that the showdown he had threatened was at hand. I believed it better not to show fear so without missing a stride I counted my options: I could turn around and go back into the building and so postpone the confrontation, I could throw him bodily off my truck, or ignore him. This last, while the most illogical, was what I elected to do. I reasoned that my first choice to run was exactly what he wanted and the second was foolish and physically dangerous and unwise as an unwitnessed event.

To ignore him was ludicrous, but that is what I chose to do! I simply walked passed him, inserted my key into the door, opened it, and got in. I briefly thought of re-locking the door from the inside, but that would have been a concession to fear, so instead I started the engine, rolled down the window, and said in as controlled and calm a tone as I could muster, "Taylor, if you don't get your ass off, you're going for a ride you won't ever forget!"

He sullenly slid off, and I drove away, trembling from anger mixed with fear.

Monday of the next week, the day before he would have been returned to MacLaren, he agreed to take his tests doing just enough that I relented and recommended that he remain in my class. This confrontation, ultimatum, and grudging concession became the pattern for our marginally tolerable relationship. Curtis had, however, so intimidated all of the other boys and many of the staff that most avoided any contact with him.

Curtis was returned to MacLaren, I don't recall why, but I was relieved. Everyday had became a clash of wills which exacted a toll on us both. But in a few months he was back. He didn't, nor did I, resume hostilities, but he made it quite clear that by-gones had not been forgotten.

I was driving the bus to Camp from the Portland bus depot every fourth Sunday of the month for the boys' home visits. The old bus had quit in Portland so I called Norm my mechanic brother-in-law (everyone should have one!) who made temporary repairs with a warning that new parts were needed so upon my return I wrote a vehicle repair request and naively assumed it would be taken care of by the next month's home visit. It wasn't.

A mile and a half east of Lee's Camp the bus stopped running. There was much turmoil as irate youths insisted they were going to get out and "thumb" a ride to their homes. Of course, I couldn't permit that and so stood by the driver's seat in the aisle to prevent anyone's leaving. I sent a reliable runner back to Lee's Camp to telephone for help. I guessed it to be an hour and a half or more before relief vehicles could arrive.

My bus full of angry delinquents became increasingly hostile. Some wanted to hitch-hike, others to walk, and the rest just angry and frustrated at our delay. I permitted them to go down the embankment one at a time to "take a leak." This distracted them for twenty minutes after which then they sat silently for another ten or fifteen minutes. Then one youth set off the clamor again with a comment about hitching a ride. I assured them that help would soon arrive.

Over an hour had passed. I was still blocking the door when I saw Curtis get up from his seat and come down the aisle toward me. He was carrying a lug wrench two feet long which he had apparently taken from the tool box under a rear seat.

"I'm leavin' this bus and ain't you or nobody goin' to stop me," he said.

I moved more to block the door, and shouted, "Taylor, you put that wrench back in the tool box now because I'm warning you if you swing it, do it good the first time because one swing is all you're going to have time for!"

"I'm gettin' off this bus an' you can't stop me!" he said in his most surly and threatening voice.

In as calm a tone as I could manage, I said, "If you leave this bus, you're going to have to hit me with that then you won't get a home visit because if I don't get you the cops will."

It was a standoff. He still held the wrench, but hadn't raised it above his waist. We stared at each other, neither spoke. Suddenly he threw the wrench onto the nearest seat, turned and ran back down the aisle to where he had been sitting put his arms across the seat back, and hid his face in his arms.

All twenty-two boys sat in stunned silence. Ten minutes later, a truck stopped behind the bus. It was Orin, the Assistant Camp Director. The boys shouted and screamed obscenities at him from the open windows. Most of these vilified me for keeping them captives on the bus. He came up two steps not quite into the bus, then, as I watched in disbelief, he turned and without a word left the bus, got in his truck and drove away!

I was puzzled. While his truck was obviously too small to carry more than two, and possibly four more in the back, why didn't he at least tell me why he was leaving or give me some idea of when he would return. I was so bewildered that I considered just giving in and permitting the inmates to get to Portland as best they could. The youths in the bus also seemed surprised too and after he left were amazingly subdued.

In another ten minutes, Doug arrived driving the carry-all, and with him the boy I'd sent for help. Behind them was Orin who talked to me as if he'd never been there before. Neither of us ever spoke of it again.

***

Curtis never came back from that home visit. I suppose that he lost so much "face" when he backed down that he couldn't return, but I did hear a lot of him from boys coming in from MacLaren. He was feared for his violent temper and accordingly spent most of his time in the "time-out" cottage smoking and playing pool. In 1991 I read of him, then thirty-seven years old, he had been shot three times in an attempted stick-up.

***

CARVER

Half a dozen visitors were standing about chatting. I was leaning against the counter a vicarious participant in their conversations, but my role was that of de facto host since I was the only staff person in the building at that moment. Our visitors were mostly counselors and parole officers, but with them were two interns who had graduated with appropriate "social work" degrees and were now training "on-the-job."

I knew all but the interns, one of whom, a young man, came over to where I was and standing and said, "Hi, Mr. Graves. Do you remember me?"

"No, I don't," I replied, "I'm sorry."

"I was here six years ago. My name is Carver . . . Marcus Carver."

"I remember the name. You were in the eleventh grade here weren't you?" I responded rather lamely, shooting in the dark hoping to hit common ground on which to continue our conversation.

"Yes," he replied. "I was seventeen, but I grew up here in just a few seconds. Is Mr. Maupin still around?"

"No," I said, "He's working as a painter at MacLaren. It pays more than running this place. I'm surprised you haven't seen him over there. But just what do you mean you grew up here?"

"The fourth time that I was in MacLaren I came to camp. I had spent most of my time at Mac' in Benson Unit. When a staff would tell me to do something, I'd say, 'Screw you.' I did that for years. I was a tough guy. Most people didn't mess with me. When I got out, I did the same thing in Roosevelt High. I was a big man, or so I thought. Then I came to camp," he continued, "and the first morning here, in the old barracks then, I guess it must be torn down now isn't it?"

I nodded affirmatively, "This building replaced it in 1969."

"Well," he continued, "I was assigned to clean the flats, sinks and mirrors as I recall. I never did my work details at Mac' . . . nobody ever made me do 'em, and finally they just quit assigning me any. So I didn't expect to do any at camp neither, but that morning as I was passing the office, Mr. Maupin called me in. 'Carver,' he said to me. 'You got your work detail done yet?' I says to him just like I had done to staff at Mac' for years, 'Screw you Man,' and then I told him, 'I don't clean nobody's shit house!'"

"Now I remember you," I said laughing. "I was in the office with Sherm that morning."

"First thing I know he threw me on my ass, grabbed my foot and dragged me down the hall to the flats. Threw me in the door and shouted, 'I'll be back in ten minutes. Better be done by then or I'll flush you down the toilet with the rest of the shit!'"

"Yes," I commented, "that's the maddest that I ever saw Mr. Maupin. You sure pushed his button!"

"Well," Carver said. "I sat there on the flats floor about fifteen seconds before I remembered I only had ten minutes to clean those sinks and mirrors so I set right to work. That was the last time in my life that I told anybody to get screwed. I decided it just wasn't tolerated around Camp, and when I got out, I discovered people treat you pretty much the way you treat them so I finished high school and then went to PSU in social work."

"You're going back over to Mac' this afternoon aren't you?" I asked.

"Yes," he replied.

"Well, Mr. Maupin would sure be glad to see you. Why don't you look him up?" I suggested.

"I don't know about that. He might still be mad," Carver said laughing, "an' I sure don't want to piss that man off again!"

***

Several weeks later, I saw Sherm and asked him if Carver had found him. He said, "Carver who?" I reminded him of the incident six years before.

"No I don't remember," he said. "It was all part of a day's work then."

"Carver remembers, that's what counts I guess," I commented.

***

When I last heard of Carver, he was a parole counsellor for the California Youth Authority in Los Angeles.

CHAPTER FIFTEEN

TOO LATE FOR GREG!

I was not surprised to see Greg in my classroom since I had been consulted prior to his coming to Camp. Greg Moss was well known in the state institutions. He had spent the first part of his life in Fairview in Salem, an institution for severely retarded children and adults.

When Greg failed to develop normally, no one suspected the two causative factors were environmental learning deprivation resulting from his total deafness. He spent the next seven years at Fairview diagnosed as severely retarded. When he was almost ten a correct diagnosis caused him to be transferred to the Oregon Deaf School where he acquired some socialization skills and for the first time started his education. Most importantly it was discovered that Greg's deafness was surgically correctable. So at age eleven, he was deemed educable and was placed in a foster home to be mainstreamed into the public schools.

Greg was still handicapped, he had missed out on the first ten or twelve years of experiential living. It was as if Greg had been placed in the sterile and protective surroundings of these institutions, and then was thrust out into a harsh environment which he little understood. He was as a newborn babe having never established systems of coping skills. Consequently he was unable to evaluate friends or to discriminate "rights" from "wrongs." He believed everything he was told and took and gave without regard for traditional ownership concerns and so was committed with some bad companions for repeated thefts.

At 16 he was placed in remedials in Lord High School at MacLaren. His reading level was tested at less than fifth grade. Greg had acquired many of the mannerisms of those with whom he had spent most of his early years. He had acquired the characteristics of the severely retarded, stammering and speaking slowly with frequent pauses and "uhhhs" as saliva drooled from his chin. Greg was so naive that he was everybody's dupe and the butt of every joke. He greeted ridicule with the fool's grin, dribbling as he stammered a response, luckily oblivious to his humiliation.

Physically he had become a "Lurch." He was somewhat stooped, large with swinging gorilla-like arms and extremely poor posture. His shoulders and arms sagged giving him the appearance of a hunch back. Greg was over six feet two inches tall and weighed more than two hundred and twenty pounds. Yet he was harassed and pressurized, physically dominated by smaller boys. It never seemed to occur to Greg that he was larger and stronger than they.

I started Greg with a simple diagnostic "word wheel" test to determine better where we should begin. I was surprised to discover that he scored well above the sixth grade level and so I continued his Individual Education Plan testing on a CAT 14C. This test showed him to be reading almost at tenth grade. His other test scores were considerably lower, attributable to his educational deprivation resulting from the five or six years his formal schooling had been delayed. All the materials that I had on hand were too difficult for Greg, but he was willing, sometimes even eager, to try, if only to please me.

He could manage simple, familiar vocabulary, but to Greg most of the terms and vocabulary in a ninth grade world geography book had as much relevance as if it had described Jupiter or Mars. The questions he asked were "off-the-wall" and so numerous as to draw much attention from his classmates to his ignorance and to get Greg started much like a three-year-old child performing for attention. The saddest of this was his total sincerity in wanting to please everyone and to learn at the same time. The heartless taunts of his classmates went unrebuffed for he didn't understand, and therefore resent, such ridicule.

Finally, I found and bought some adult basic education workbooks which emphasized real situations in auto and appliance repair and another was the ongoing story of three construction workers traveling from job to job around the nation. Not only did these books emphasize adult learning situations, but they became classroom "best sellers." Everyone wanted to read "what Greg is." Thus it was that he attained a certain degree of respect and acceptance from his classmates. These Cambridge books provided from primary through fifth grade reading levels in a vocabulary more within Greg's range of experience. I had hoped that they would help him identify with "working stiffs" and prepare him to someday earn a living.

Still outside of the classroom he was "put down" by everyone and was socially rejected. He was at the very bottom of the camp's pecking order which was as formal and rigid as in any chicken yard. All this changed one holiday when instead of school we were having informal recreation such as pool, ping pong and board games.

"Hey Mr. Graves will you come out and referee a game of football for us?" Jay Harper asked. There had been some injuries from rough football so the group was only allowed to play with a staff member setting and enforcing rules.

"Yeah, yeah, good idea. How about it?" several others chimed.

"Okay," I said, asking the other holiday staff if he wanted to join us or take charge of those remaining inside the building.

"Okay if I take the ball out?" Jay Harper asked.

"Take it. Everyone who wants to play football come on outside now." I shouted.

Outside I counted thirteen. The wrong number to play six on a side. This problem was quickly resolved when Greg stated that he didn't want to play, not surprisingly since he had never played before and he was likely aware that others would be "laying" for him trying to "cream" him. I was relieved when he said he wasn't going to play.

"Can I stay out, Mr. Graves?" Greg asked indicating that he didn't want to go back into the building.

"Sure," I said. "You can walk the sidelines with me and stand where I tell you as a down marker."

Then to the others I shouted, "Gather around." When they were in a semi-circle I explained, "Here's the rules. No tackle." I passed out pieces of cloth. "Put these flags in your belts or tuck them into the waistband of your pants. Ends have to be free, don't tie them or put them in your pockets. You may block on the line of scrimmage, pass block, and block tacklers above the waist, but the ball carrier may be stopped only by pulling out one or both of his flags. Everyone understand the rules? You cannot use your hands except for passing and catching, okay?"

The two sides were drawn, not by lot but by directive from Jay Hargis, the camp bully and loud mouth. Jay was a leader who was respected for his loudness and his bluff and his bluster. He assigned boys to one team or the other according to their known prowess on the field of play. And of course, the best players were on Jay's team. This is how the game began.

Jay was, as one might logically expect, the quarterback team captain, and coach. Right away he threw for a touchdown, then he carried the ball fifty yards for another. Soon his team was four TD's ahead. But then one of Jay's teammates, a house boy, was called in to help prepare the evening meal so they needed another player.

"Hey Greg, get in here. We need you to make the teams even," Jay shouted at my "yard marker."

Greg dutifully shuffled off to join Jay's team, but upon arrival he was immediately traded to the weaker team by Jay (who was also the Director of Player Personnel). The game resumed with Greg looking bewildered. On the next play Jay ran right by him.

"Stop him at the line Greg. You can block Jay, just get in his way that's all you have to do," Greg's teammates implored.

A play or two later, when Jay ran at Greg, he was stopped cold, and very nearly knocked unconscious. Greg was learning the game, and with his size, he was a natural. Once more he knocked Jay backward onto his backsides causing him to jump up waving his arms wildly.

"Stop the game! Stop the game!" Jay called. "Greg, you come over here on this team. It's plain to see that you don't know nothin' about football, so it ain't fair that those guys have to have you on their team."

"It's okay, Jay," one of the other team, "We'll teach him!"

"Yeah, Jay I wanta stay on this team," Greg added.

"Well, it's time to quit anyway," said Jay, heading for the building with the football under his arm.

This event earned Greg respect as a physically potent force, but he didn't stay at camp very long and was placed in a group living unit as a full time worker. He was nineteen years old and it seemed that school was less important than for him to learn to support himself and learn to make economic decisions.

***

Greg came back to visit us once on the Greyhound bus with fifty dollars stuffed in a back pocket of his Levis. He was wearing a large black Stetson type hat and cowboy boots symbolizing his new life at a Bend group home ranch. He had come the last two miles from Tillamook by taxi. The rest of that day he was in and out, several times he asked to use the Camp telephone to call for a taxi. Each time, I thought that he was leaving, but soon he would return, always by taxi.

"Mr. Graves, some buddy's stealin' my money," he said.

"Greg, I've seen you stuff money into your hip pocket, chances are that it just fell out."

"Naw, I had fifty dollars when I got to Tillamook, now I only got five left."

"How many times have you been back and forth in that taxi?" I asked.

"Well, I dunno, maybe five times," he replied.

"And how much does it cost you each time?" I asked.

"Three dollars and sixty-five cents, I think," he responded, "but, let's see now that's less than twenty dollars I spent so where's the rest of my money?"

"Round trip," I said, "The taxi driver must charge you going and coming."

"Oh, I guess so," he said.

"You've got enough to go back to catch your bus so it all worked out okay, didn't it," I commented.

"Yeah, that's the good part ain't it? Just like on tv when Aunt Bea always says 'every cloud has a silver linin', an' my silver linin's in my pocket right here. Oh! Oh! you know what, Mr. Graves? I just bought cigarettes from Billy and so I can't catch my bus so I guess I'll just have to stay here with you."

"Greg, it'll be my pleasure to drive you to the bus depot!"

***

About two years later I read that an unidentified male had been discovered in a ditch beside a highway in Southern Oregon. Oddly the paper reported that the person had been wearing a World War II era dress blue sailor's uniform. Several weeks later I learned that the corpse had been identified as Greg who had died of exposure resulting from a drug overdose induced coma.

CHAPTER SIXTEEN

BJ

I had saved the tall brown haired youth for the last of the new arrivals to interview. I sensed this could take a long time. After getting the first two started, one on his pre-tests and one off to Mr. Hogland's class, I asked the tall boy to sit in the chair facing me beside my desk which I kept there for just that purpose. He sat down as requested and I introduced myself as his camp teacher for social studies and English classes.

"So let's see what classes you need to take."

"I'm a senior, Sir, and I only need a credit and a half to graduate."

I scanned his transcripts and replied, "Maybe so, twelve and a half, and two more make fourteen and a half and nineteen and a half . . . and yes, you're correct. How did you know?"

"My parole officer had my school counselor figure them before I left Condon."

"Who is your P.O.?" I asked.

"Mr. Hamilton," he answered.

"Did he say what he expected you to accomplish in school?"

"He wants me to graduate from high school."

"Well, you should be able to by the Christmas home visit," I advised him. It was then mid-October. "Okay you need a half in English, and one in social studies—how about consumer economics, and life skills?" I asked, seeing a notation from Mr.

Hamilton that he was a new father who would need to get a job when he was released.

"I see that you have just had a new baby."

"No Sir, my girl friend," he was emphatic, "Desiree did!"

"Well, that's what I meant to say. You did have a degree of involvement though, right?"

"What?" he asked.

"You are the child's parent?" I inquired.

"No Sir, I'm not married. Desiree and I are just friends."

"Good friends I hope. Mr. Hamilton has made a note here that you are going to marry her when you get paroled."

"Well we're going to live together at least," he responded, then added emphatically, "I sure love that little lady!"

"So you'll marry her then you'll become a parent?" I laughed.

"Wah'l, I don't know. She's on ADC now."

"ADC is Aid to Dependent Children isn't it?" I asked. "She gets money from the state, right?" I said, thinking more and more that I knew this youth from somewhere.

"Yes Sir, that's what it is, and Desiree says she won't marry any man who cain't provide better for her than the ADC does."

"Well, that certainly shouldn't be difficult. ADC probably doesn't give her much money, especially with just one child. You know ADC pays for each child you have," I said.

B.J. looked at me intently and inquired, "How much?"

"Oh my God! This kid's thinking about going into stud." Then it came to me, where I'd met him, "Mayberry RFD's, Gomer Pyle!"

"No, no, I mean when you get your diploma, and learn how to fill out job applications and tax forms, and all of that, then you'll be able to earn lots of money. Maybe working in a lumber mill or on a ranch, or flying a 707! Who knows where an education will get you. Right?" I tried to be enthusiastic, but some of it, the work and taxes, were hard even for me to get excited about.

B.J. was surprisingly diligent. His work was done methodically. He read at a snail's pace and wrote even slower. Many years before I had had a student very much like B.J. who got an appointment to Annapolis so I wasn't discouraged. Neither was B.J. We enjoyed our association, and he spent many minutes each day—even hours—in the tutorial chair by my desk. His one great failing was "flats call." "Flats" were what inmates called the rest rooms. Well it seems that B.J. enjoyed the sit-down solitude of the flats. Since this facility was just across the day room from the classroom I allowed students unrestricted access to it. B.J. established an early reputation for spending so much time there that I had to limit him to two visits a day.

There were so many wonders in this world that B.J. was discovering. In his "Gomer-ish" way he could get excited learning simple things I thought everybody knew such as that in pregnancy the mother nourishes her child though an umbilical cord. He was so astonished that I expected him to say, "Well Gollleeee!" like Gomer. Actually he just said, "Don't that beat all!" Well, of course all this additional learning had nothing to do with senior English, life skills, or consumer economics. I didn't mind so long as he was learning for it seemed he wanted to know about everything. That was fine. I assumed that everyone else would be as pleased as I about B.J.'s wide-ranging, liberalized education. Not so, some people were waiting impatiently for him to be released. They lined up sort of as follows: Desiree. Desiree's parents and their Marryin' Dan preacher man, B.J.'s counsellor, Mr. Hamilton, and the administration who wanted more room for other youths in need of educations. As the weeks turned into months I began to get more and more inquiries about B.J.'s progress.

"He's doing a lot of paper work!" I would answer with a grin.

One day in consumer economics, B.J. discovered the "Great Society" concept. He ran across a pamphlet called "How Your Government Helps People." He took an immediate interest and wanted to find out more. This was much nearer our course of study than umbilical cords so how could I refuse? I sent for many publications from various departments of welfare and public assistance. We made field trips to government offices such as the Employment Division (which most people especially B.J. commonly refer to as the unemployment office), and to Family Services. We learned how to qualify for food stamps, free school breakfasts and lunches, bus passes, and even fishing licenses!

Then finally the great day arrived. B.J. graduated the first of May. He had taken six and a half months to earn one and a half credits. Still I was proud of him for I knew he'd learned his lessons well and had it in him to be successful in whatever line of endeavor he chose.

I got out the graduation hat, an old black mortar board with tassel and all, asked the cook to bake him a cake, and at lunch called on him for words of inspiration. All he could say is, "Well now, don't this beat all!" Then lacking the necessary canons to give B.J. the salute he deserved, we gathered in his beloved flats and stood at attention for a two urinal, three stool flush in B.J.'s honor.

***

Several years later when I saw Mr. Hamilton I ask about B.J. "Yes," I was told. "He's a good citizen, a father, a cub master, even PTA President, but he hasn't done an hour's work since he left camp. He knows every source of assistance the state and feds offer and uses them all to support himself and his family."

CHAPTER SEVENTEEN

A LONG THREE HOURS

"Mr. Graves, Mr. Graves, Maurice has gone crazy and is tearing up the dorm!" a boy who had been "on flats call", an institutional jargonism for "using the restroom", came running into my classroom breathlessly to announce.

At that same instant I could hear crashing and shattering sounds from the large dormitory room only twenty feet away. The half-dozen students in my classroom rose in unison and headed for the door to see the excitement. Purposefully, I had placed my desk near the doorway so I would be able to bodily block their exit. I demanded that they take their seats and stay put while I investigated the disturbance. Then at a pace just short of a dash I headed for the dorm.

As I passed Mr. Hogland's classroom I looked in to see him occupied with something on his desk and his students apparently were oblivious to both the commotion and the youth's report of it. At the doorway of the fifty foot rows of beds I could see Maurice at the far end tipping over the iron framed bunk beds and throwing night stands about. As I neared him, he grabbed two handfuls of curtains from several windows jerking them off their rods and throwing them angrily to the floor.

Maurice had chosen not to go to work that day by the customary means of feigning illness. This was not unusual for new boys reluctant to do the sort of physical work which was required by the State Parks work crew. Maurice was new. This was my greatest concern. Not knowing him, I didn't know what to expect from him or what hazard he presented. What little I did know was bad! He had come from Dammasch Hospital where he had been held for psychiatric diagnosis. This probably indicated a history of violent behaviors. "How extreme?" I wondered.

I was not frightened although he was a tall strong youth who could likely have overpowered me, and certainly could well have thrown something at me or hit me with some of the rubble he had just created. I hesitated only briefly, for I had learned long before, hesitation may often be interpreted as weakness in such crises. I had to do something! Doug Hogland and I were the only ones in camp. All of the other staff including the Camp Director and Counselor were gone, and would he until four p.m. It was now just past one in the afternoon. I didn't want to call on Doug to help me because one of us was needed to maintain whatever calm was left in the classrooms.

I knew that I couldn't physically subdue Maurice should he resist, and besides this was a gamble not worth taking so I shouted in as authoritarian a tone as I could command under the circumstances, "Maurice, you stop this immediately." And then screamed, "You hear me!" He stopped, sat on a bed, and began to sob. His shoulders heaved and he sat on one of the few still upright bottom bunks. His arms were folded across his knees. Maurice was broken, a pitiable child in a man's body. I sat down beside him and put my arm across his shoulders and neck patting his shoulder with my hand. It seemed the right thing to do at that moment. Because he was too emotionally shaken to speak we just sat there for two or three minutes silently.

When he had calmed sufficiently, I said to him, "Maurice come on down to the office with me where we can talk." I probably meant, ". . . to the office where I can talk," for the office was my "turf" where I knew I could command the telephone and be visible to Doug should Maurice again become violent. In the office too were the handcuffs and restraint straps should I need them.

He came with me, nearly bent double and sobbing, still unable to talk. As we entered the open area between the classroom and the office, we had the usual audience of curiosity seekers. I really didn't expect my students to stay in their seats when I suddenly left, for they never do, but at least to keep them in the classroom. They were there, peering through the windows and hanging out the door. Doug left his desk and sent them back to their seats, but not before they had snickered and pointed accusatory fingers at Maurice who thanks to his arrogance and prior incidents of physical intimidation was not well-liked. They showed little sympathy for Maurice and he, surprisingly, interrupted his trance-like grief state long enough to snarl vicious unpleasantries to them: "I'll get you, you 'mothers'!"

I was surprised that one so emotionally destroyed could instantaneously rouse himself to such hostility, but deceived by his docility, I was not forewarned. Maurice and I went into the office where I asked him to take a seat. Since Doug and I were the only staff in the building, I deliberated what to do. I had to get him handcuffed for his safety and ours, but how? If I called Doug to back me up and tried to physically force the cuffs on him, our gallery of gawkers could not only cause him to resist, but one couldn't be sure that a riot might not ensue. So I had to talk Maurice into letting me handcuff him. I knew that wouldn't be easy.

"Tell me what's bothering you, Maurice," I started.

"I don't wanta be here," he said. "I was sent 'gainst my will. I wanta go back to Dammasch."

"What do you like about Dammasch?" I inquired, maneuvering for an opening to ingratiate myself to him.

"Freedom," he replied. "It ain't like here. There's no work and no school . . . and girls—I had a girl friend there."

"What did you do all day?" I asked.

"Well, we was locked up most the time, but in a big recreation room so I played pool and watched tv and played cards and . . ." He looked toward the classrooms and muttered, "Bastards!"

"Don't pay any attention to them," I said, realizing I had momentarily lost rapport. "They'll do anything to break up the monotony of school. Getting you to go 'off' on them is today's entertainment."

"Yeah, me too," he replied, at last some common ground with his camp mates. He added, "I hate school too."

And so we talked, for a half hour. He told me of family strife which had separated his parents. His father was some kind of religious minister who had deserted his mother for another woman and he said that he hated his father. I empathized, all the time working my way toward getting handcuffs on him.

"You know, Maurice, you have to go back to MacLaren for wrecking the dorm," I ventured boldly.

"I want to go back to Dammasch, not Mac'," he responded.

"Well, I'm sure that when the psychologist in the clinic, at MacLaren finds out how badly you want to go, he'll send you back to Dammasch," I assured him, but knowing that once Dammasch had rid themselves of a dangerous delinquent patient such as Maurice not even a court order could force them to re-admit him.

"Well all right then, put the 'cuffs on me and I'll go if you'll take me right now."

"Sure," I said, "I'll have to use restraints too."

"Okay," he said.

I opened the wall cabinet in which the straps and cuffs were kept and took out the restraints which hold an inmates hands at waist level either in front of him or behind him.

"No cuffs! I need those cuffs where the hell are they? Probably they weren't put away after the hiker capture the other night."

I rummaged around in drawers, and the individual staff's mail boxes, finally calling Doug to help me look for them.

"I'm sorry, Maurice, but you'll have to wait until the evening staff comes on duty. He must have the cuffs put away somewhere. I don't know where, and I can't take you without them."

"I wanta go now. I'll go with you without being cuffed, and I promise not to run. Okay?"

Now I had decided to try to stall him for two hours until two more staff would come to work so that Maurice could be restrained and taken to MacLaren. I believed that I could use up an hour or more by stalling my way through the normally tedious procedure of checking-out his personal belongings.

"Well then let's go get your stuff and get packed to go," I opened the office door motioning him across the day room to Doug's classroom in the back of which was the personal clothing lock-up room. He walked ahead of me very dejectedly.

There were a half dozen youths in the classroom and Doug was seated at a desk by the door.

"Hey ass hole! What'cha cryin' about?" and other muttered comments set Maurice off. Wordlessly he leaped across a table, grabbed one tormentor by the neck shook him like a dog does a cat against the wall, then he hit another, and grabbed a book to throw. I seized his arm from behind and getting a firm grip was able to restrain him from any further attacks.

"Come on Maurice, let's go back to the office, okay?"

He came with me, once again sobbing and muttering threats against those who had enraged him. In the office we resumed our time consuming counselling session, I was still buying time I thought waiting for someone to arrive with handcuffs. I thought about calling the sheriff's office to send out an officer to bring handcuffs, but because of his violent behavior I was afraid to make such a call which could only incite him into another outburst.

"I wanta call my mother," he sobbed. "I need to talk to her."

I thought about it for a moment and said, "Okay, what's her number?" He gave me a number which I dialed.

I handed him the telephone. He waited, phone in hand, but not to his ear. I could hear a squeaky male phone voice saying, "Who's there? Who's there?" Then Maurice held the telephone to his mouth and began screaming obscenities and profanities to whomever was on the line finally throwing the telephone violently onto the desk.

I picked it up and asked, "Is this Maurice Murray's mother?"

"No," a male voice answered, "this is his father, and what do you mean to let him call and talk to me that way? And where is he? And who are you? And . . .?" I just hung up the receiver, feeling more helpless than ever . . . and mad as hell at whoever hadn't put the handcuffs back where they were ordinarily kept!

"I wanta go now, an' if you don't take me back to Dammasch I'm going to 'go off' on you 'cause I can't control myself anymore, you hear me, Mr. Graves?"

Yes I heard him, but after trying to keep the lid on now for over two hours, I was either going to give in or try to take him to the county jail, but I had to do something. It was clear that I had stalled as long as I could.

"Okay I'll take you without handcuffs. Let's go!" I took a vehicle key out of the key cabinet.

"I gotta get my stuff. I ain't going to leave without my clothes!" Maurice said. "Them 'mothers' in there'll steal 'em fer sure."

"If I take you back through the Mr. Hogland's classroom are you going to 'flip-out' again?" I asked, back to stalling for time. I now hoped to use the remaining twenty-five minutes until Cecil Omans, who was more often early than late, would arrive at work and hopefully show me where the handcuffs could be found.

"No, I promise I won't 'go off again'," Maurice answered.

"Okay, let's go," I said, hoping that this time Maurice wouldn't assault anyone.

It was a mistake. This time Maurice went berserk without provocation. He picked up a chair and hurled it at a boy who ducked under the table to narrowly avoid being hit. Maurice had just picked up another chair as Doug confronted him.

"Put down that chair kid," Doug ordered sternly. I knew Doug's tone of voice Maurice had better obey. Doug was a huge man, six feet three inches tall and two hundred and seventy pounds who had played six years in the NFL. He was angry.

"Screw you, mother ——er," Maurice foolishly responded. Doug only hit him once, squarely in the center of his face. Maurice reeled backward and fell against the wainscotting under a bulletin board conscious but stunned as thin lines of blood began to dribble from each of his nostrils.

In the doorway stood Cecil Oman.

"Cecil, I've been waiting all afternoon for you," I exclaimed. "Where are the handcuffs? We needed them in the worst way and I couldn't find them."

"I just brought in three new sets day before yesterday," he replied. "Have you had a problem?"

"Yes," I said sarcastically. "We've had a problem."

I took Maurice into the office once more. Cecil opened the safe and under several manilla envelopes and some loose papers in three neat little boxes marked "Peerless" were the handcuffs which I had been frantically searching for all afternoon.

"Here they are, nice and new, shiny aren't they?" Cecil asked proudly opening a box and handing me a pair of cuffs.

"They would have been a lot prettier about one-thirty this afternoon," I commented wryly. "Let's hang 'em up in the key cabinet in plain sight from now on, okay?"

***

I checked Maurice's rap sheets and psychological reports to discover how emotionally disturbed he really was. He was reported to have established a pattern of control as early as the primary grades in which he would behave irrationally, threaten, and assault classmates. Well meaning teachers for the most part responded just as I had, and in the end, Maurice usually got what he wanted, special treatment, isolation, and attention.

Maurice was taken back to MacLaren, not to Dammasch. I didn't see him again for four months because I took the summer off to build a new house. When I hired several camp boys sight unseen tn help unload lumber, there was Maurice. I asked Doug how Maurice was getting along in Camp this time.

"Just fine," Doug said! "and he goes to work every day he's supposed to. He hasn't taken even one '24'."

Kitchen and dining room at Camp Tillamook;
cook Erma Blackmore working at rear.

CHAPTER EIGHTEEN

THE DEVIL'S ADVOCATE

A Tuesday afternoon, after class dismissal, I was sitting in the classroom when I heard voices of the boys and staff who had just come in from MacLaren. I had very little curiosity about our new arrivals because I would meet them soon enough in class. But then Dan, a returnee, came in to tell me that he had come back after having been at MacLaren the mandatory sixty days "dead time", his penalty for trying to bring marijuana into Camp.

"Hi, Mr. Graves. I'm back," he said.

"So I see," I replied, "I hope you've learned a lesson."

"Yeah," he responded unconvincingly.

It was apparent that bringing up the reason for his removal from the camp program was an unwelcome topic of conversation.

"Who's your friend?" I asked noticing another youth whom I had not met standing behind him.

"Todd Amos," he said.

"Hi Todd," I responded, fascinated by Todd's flamboyant style of dress. Over faded blue jeans he wore crotchless leather trousers that were fringed along their outer seems, and above those a leather vest inscribed with many randomly placed geometric designs which I assumed were native-American in origin. Todd wore no shirt under the vest, but his hairless chest was draped with multi-colored ropes of necklaces composed of many shaped beads. Atop his unkept shoulder length blonde hair was a coonskin cap, furry tail and all! On his feet were buckskin fringed moccasins. Todd's hair and clothing marked him as having come to us from Hillcrest School where students were permitted much more "self-expressive" styles of dress than were the boys at MacLaren.

"Trapping beavers?" I asked, chuckling at my own humor as I sarcastically belittled the inappropriateness of his attire.

"No," he said, either ignoring or not understanding my ridicule, "I'm a warlock."

I knew warlocks were witches, but that really didn't seem in keeping with his costuming so I responded, "A warlock on the warpath? Not taking scalps I hope."

It was apparent that he was annoyed by my jokes, and since my curiosity had now been piqued, I inquired in a more serious tone, "What is a warlock?"

"A warlock is a male witch," Todd replied.

"Oh?" I said. "What do male witches do?"

"The same thing female witches do. Work spells and heal sick people and call up spirits and . . ."

"You can do all that?" I interrupted him. By now two other boys had arrived and were listening. It seemed that Todd responded theatrically to the larger audience, making me more determined to discredit him and so prevent the establishment of cults in camp.

Several years before, psychologists' opinions had held that it was more wholesome to integrate the sexes in juvenile residential facilities. Hillcrest School was chosen to become co-educational. At the time, I had predicted that such a policy would lead to sexual promiscuity and so eventually necessitate re-establishment of segregation of the sexes. In a year or two it became apparent that I had been wrong. The males were not seducing the females, quite the opposite; the girls feminized their male cohabitants. Boys began to refer to each other in what I had always thought of as uniquely femininely verbally assaultive names such a "bitch" and "whore." Machoism had eroded though. Hillcrest males were much more likely to wear jewelry and perfumes, to shave their legs and underarms, and to wear nylon bikini briefs (All of which were also in vogue as a part of one of the greater society's acclaimed unisex style movement, and so were not uniquely a Hillcrest phenomenon).

"No," Todd retorted, "I can't do all of that. I'm still learning. I'm an apprentice sorcerer."

"Who do you apprentice to, or under?" I asked, then adding, "I hope you don't mind all my questions, but you're the first apprentice sorcerer I've met in person, and I want to learn all I can. It may be a new career opportunity for me." I winked broadly at our audience.

"The Devil . . ." he said, looked directly at me and repeated slowly as if for affect, "I'm apprenticed to the Devil."

"Wow!" I said, "I work for Mr. White, and that can be tough! What Devil do you work for?"

"The anti-Christ," he replied.

"Then you must be an admirer of Hitler and people like that," I said, probing for a weak spot, a point of ignorance to contradict him, but he was intelligent and very well informed. I later learned that prompted by "freedom of information" decisions and inmate demand, Hillcrest had acquired dozens of books for their library about mystical and occult subjects.

"No, Hitler was an evil man, a politician, stupid but not a warlock. Charles Manson is one of us," Todd replied.

"He's 'fried' his brain on LSD and peyote," I noted a comment from one of the bystanders with some satisfaction. It seemed that I might be making headway in opposition to Todd's proselytizing.

Todd ignored him and continued, "I'm a warlock of a coven of witches. Dana Thorogood, and I had a coven of seventeen at Hillcrest. Dana's there so I've come to start a new coven here."

I looked around the room at several more interested onlookers who had joined our group.

I said, "At Hillcrest when and where do you meet?"

"After group," he said, "and Wednesday nights."

Group I knew meant boys and girls involved in the Group Guided Interaction Program which Hillcrest had utilized for many years. I said, "Staff permits your coven to meet?"

"Staff doesn't have anything to say about it. Group makes that decision," Todd replied. "There are other covens in other groups . . . most groups have them."

"Yeah, I heard Dana took over your coven and group voted against you," said one of the bystanding youths who was apparently from Hillcrest and knew of what Todd was telling us about.

"They won't let us do witchcraft here," another said.

"I'm sure that we won't," I said. "Personally I think it's garbage. I don't believe in covens, witches, or devils. I'm sure Mr. White won't allow nutty groups of this kind in camp. Also you're going to have to take off those silly clothes and jewelry because your work supervisors won't allow you to wear them for safety reasons . . . even your hair will have to be cut shorter or worn in a hair net to keep it out of chippers and other machines you may need to use on work crew."

"Nobody can make me dress any other way than what I want," Todd said. "The U.S. Constitution says we have to be allowed to have clubs and organizations and meetings. We learned that at Hillcrest. So I will dress however I want and start a coven here if I want. No one can stop me."

"Hey man, put a hex on him!" said one bystander, pointing my direction and obviously referring to me as "him."

"Yeah," said another, "sic the Devil on Mr. Graves."

I was reminded of the short story "The Devil and Daniel Webster" in which Webster debated the Devil and won freedom from damnation for a soul. Todd and I had an audience of several souls we were striving for. Since ridicule and sarcasm didn't seem to be successful tactics, I decided to try another approach to probe for weaknesses in Todd's logic.

"Why do you wear buckskin clothing? Does the Devil make you do it?" I said, attempting to counter his outburst and to draw attention away from the commonly misunderstood, over simplified Constitutional open-meetings, free speech rights argument then so prevalent among inmates as to be considered irrefutable.

"I don't wear artificial fabric, natural leather is the best because it was once a living thing. Wool is good, and linen and cotton are okay."

"What do all the symbols mean?" I inquired. "Any of them a rank, like private or corporal maybe?"

No one laughed.

"This one," he said, "is the nearest thing to that." He pointed on his vest to what seemed to be a crude drawing of cattle horns in a U-shape with a down-thrust handle resembling a music tuning fork. "Then this one," he said, "shows my allegiance to the occult." He went to the chalkboard and deftly drew intermeshed triangles forming an six-pointed star. "This," he said, and drew a Christian cross, "is the sign of our enemy. Most of these," he said pointing at symbols, "are secret. Some are learned in the coven and some, only I, a warlock, know . . . the rest are mystical, nobody knows what they mean."

"Here's one," I said pointing, "that you could play tic-tac-toe on if your coven meetings get boring."

No one laughed. Then we were interrupted by Floyd Case the afternoon staff who called, "All you new boys come to the laundry room now to get your work clothes for tomorrow's work crew."

"Well Todd, I don't think your witchcraft is going to catch on here. I'm sure that these guys," I waved my arm in the general direction of the five others, "aren't going to be taken in by such silly nonsense."

"I'm deadly serious. We'll see who wins out . . ." Todd paused as if to say something more, but then left causing a soft rustle and tinkle of bells.

"We'll put you in some decent clothes, get rid of all those necklaces and rings, and make a real man of you here at Camp," I heard Floyd tell Todd as they entered the flats.

***

A few days later when I came to work, Chris, the night staff, said, "Did you hear what happened last night?"

"No," I said.

"Todd and three other kids ran away, stole a VW bug, and headed for Portland on the Wilson River Highway. Fifteen miles out of Tillamook, at a high rate of speed—well, as fast as a bug will go—skidded in front of an oncoming chip truck, two were killed."

"What about Todd?" I asked.

"Todd was killed," he said, paused and added, "instantly. Dan died on the way to the hospital and Pat's in a coma, not expected to recover consciousness. Joe's the lucky one, not even a scratch!"

I was aghast, "Did Joe say why they did it?"

"Todd was their leader. He didn't like having his necklaces and leathers taken away, and refused to go on work crew. The others were influenced by him I guess, all but Joe who wasn't really one of them. He said he just decided on a whim to go with them as they were leaving the building."

I said, "Todd hadn't started school yet, but I talked to him enough the day he arrived to guess something bad could happen."

He continued, "Todd was driving fast and reciting a gibberish he said would make him immortal. He drove faster and faster saying over and over 'Nothing can kill me, nothing can kill me, nothing can kill me'. Joe told us, by the time he realized how crazy Todd really was, it was too late to get out of the car! He said they all pleaded with Todd to stop and let them out, but he wouldn't. What do you think of that?"

I was asking myself whether or not I had contributed to this tragedy by rejecting Todd's ideas, was I too critical, was it Todd's constitutional right to dress as he chose and practice witchcraft if he wanted? Was anyone to blame? Could the accident have been pre-ordained—God's eternal combat with the Devil? Irrational? Maybe, but I couldn't hope to explain why I felt that in death Todd had made a better case for the triumph of evil over good than he had in life.

In the Camp Tillamook office are, left to right: Floyd Case, GLC; Erma Blackmore, cook; and (holding paper) Dick White, Camp Director.

THE GANGSTER ERA

The gangster is an outgrowth of neglect. These youths are the natural consequence of broken, drug addicted homes, and of the filth and moral decay of the inner city. Drugs, especially crack cocaine, are the fuel that drives the gangs. Due to the widespread notoriety of gangsterism, many and perhaps more dangerous clones such as white supremacist and Asian and Hispanic gangs have evolved. Today's gangsters are crudely patterned after the mob to control vice and narcotics on their "turf." Nearly all gangsters are sociopathic illiterates who seem likely to decimate each others' ranks through murders and subsequent imprisonments.

The original marks of the gangster were expensive sneakers of a half dozen brand names, "starter" athletic jackets, and colored bandannas or "doo-rags" worn on the head, neck, or in a pocket. The gangs' leaders may be identified more by expensive leather coats and Italian shoes. These gangsters, unlike those of film fame, prey on each other and inner city residents who are among the poorest people in the world so most succumb in poverty and degradation to their own vices and ignorance.

CHAPTER NINETEEN

BIG AL

Albert was the first black gangster at camp, maybe the first anywhere. He was a magnificent physical presence. When Albert strode across a crowded room inmates stared and stepped aside. He was feared; he knew it, and he gloried in it, but Albert also had an intellectual side which unfortunately he never really utilized but was his undoing.

He stood an inch or two over six feet with a broad face and an aquiline nose and straightened hair. He was very muscular with wide shoulders and huge thighs and calves. His half-brother was a pro football player. I was confident that Albert could have been a great player too, but that wasn't his ambition. I don't think he ever probably held a football in his hands, for his ambition was to be measured in blood and tears . . . not sweat.

I have no idea whether Albert used drugs. I don't recall any instances of his smuggling or using them at camp. He did want to become a drug czar in the back of a large black limo surrounded by henchmen and body guards. I know because he told me so.

He wasn't the kind of student that I liked to sit and make small talk with. He wasn't fun to talk with, instead conversation with Albert was an ordeal. He knew about intimidation and used it. Albert had a fierce glare which looked as though he might begin beating on his chest, if not somebody else's, momentarily.

I never saw or heard of him assaulting anybody. He didn't have to. Fear was a stronger and safer weapon than assault. He knew institutional rules and knew just how far to push boys and staff. He wanted it understood that he could beat anybody at anything any time. "Step aside Man 'cause here come big Albert!"

Albert just couldn't beat me at ping-pong. We played the institutional style game which oddly enough seems to be used all over the country in correctional facilities. It differs in that the serve changed every point and one had to win by two points. I had perfected trick shots, some with forward spin so as not to bounce high and others with side and back spin. I played Albert by "dink and slam" forcing him to come to the ball, then when he did I'd hit long and on his arching return, slam it into the table and over his head. "Hey Man, you makin' a fool of me or what?" Most of the time he was a good loser, but sometimes he would destroy the fragile celluloid ball with one last huge frustrated smash. Then I had to restrict him from ping pong for a week which only made him angrier and meant I had to play less challenging opponents.

But that's how we came to respect each other. Physically he could have crushed me like a peanut. His reaction and coordination across the table from me was wonderful to see which only made me the more determined to win. Albert matched my competitiveness.

In class, Albert was intelligent. He talked of going to college like his brother had, and he could have if he'd wanted for money was not a problem. His father and step-mother were well-to-do. His father had funeral parlors in six eastern cities and his mother, a beautician, owned several shops. He used to joke that his mother "pruned 'em and his father planted 'em." Albert didn't care about money; he wanted to be feared and respected much more.

Albert's great obsession was becoming a gangster. He didn't mean like the "wannabes" who drive by and "gang-bang." In fact they didn't exist yet. Albert meant like the Mafia,

The Mob. This was the era of black power, "black is beautiful." Militancy, power, and equality in all things, even in organized crime!

"Why," he asked, "do whites get all the big money rackets, like pension funds and corporate shake downs while us blacks work shell games and sell white mob heroin and cocaine to each other?"

In the rising tide of equality, his reasoning was sound, even if warped legally and morally. "Why didn't the white dominated police cut off the supply of illegal drugs to blacks. Were they on the take? If a black could be a janitor or a fry cook, why not a mob boss?" he asked. I couldn't answer his questions. I didn't know. I conceded that some policemen were on the take, but counter-questioned, "Is black society culturally better off if it has gangsters, or do black mobsters degrade their community even more than outsider corporate crime. Why let us 'honkies' sell the black man shoes, autos, or tv's? Walmart and R.J. Reynolds, that's where the really big money is! Legitimacy is profitable."

"I'm goin' to Detroit, and start up my own Mafia, a black Mafia. It's time us blacks profited from crime like Whities do."

"It's a crazy way to die," I replied, "The big mob guys will only let you live until you get in their way. When you start cutting into their profits, they'll kill you."

"All depends on who shoots first don't it?" he countered.

"If you want to do something for your race be a Martin Luther King or Medger Evers or," I hesitated then added, "even Malcomb X."

"Man!" he laughed, "It's safer bein' a gangster!"

I was really glad to see Albert go when he left camp for two reasons. His intimidations didn't make my job any easier, and he was getting better and better on the ping pong table. It would have been only a matter of a few weeks until he would have beaten me every game.

Less than a year after Albert left camp he was found in an alley in Detroit with his hands tied behind his back and a single bullet hole in the back of his head.

CHAPTER TWENTY

THE SELF-MADE MAN

Robbie's fame as the greatest snatcher of citizens-band radios in K-Falls preceded his coming to camp. CB's were then relatively new on the market and quite expensive. A good one could cost five hundred dollars and this kid had led a gang of thieves who specialized in stealing CB's, removing any identifying serial numbers, then selling them "hot" and cheap to trucking firms and independent truckers. Robbie was a "big man" on campus at Woodburn. How much was true and how much Robbie made up was difficult to say. My informants snickered as they told me how he had stolen radios from one trucking company and then had sold them back to the same company—not just once but several times, "Pretty smart dude, huh?" When I asked where they had gotten their information, they always said, "Robbie. Robbie told us how he done it. Pretty cool operator isn't he?"

It was common then for kids going to MacLaren for medical or dental work to be temporarily lodged an hour or two, sometimes even a half-day, in Thayer which was one of the two cottages that had staff to supervise inmates during school hours. This created a great cesspool of misinformation and was fertile ground for those shirkers who had connived to escape the tedium of campus school. This fact alone was a tribute to their deceit, and so kids like Robbie groomed their reputation across a table littered with playing cards or while watching daytime reruns of Perry Mason even before they arrived at Camp.

"So," I asked, "What makes you think this Robbie will be approved to come to camp . . . and if he is, why this camp? He could be sent to one of the others, or Picture House in Portland maybe?"

"Robbie will be coming. He told us so, and he's so smart that he does just what he wants. He 'chumps' staff and counsellors at Mac' all'a time. Oh, He'll be here all right!"

I was skeptical. The Superintendent was Dick Peterson who had come up through the ranks at MacLaren and the camps. He knew the camps, the staff, teachers, and counselors, and remarkably most of the inmates personally. Mr. Peterson had instituted an "earn your way through the system and out" program in which release was contingent upon the attainment of behavioral, work and school goals so it seemed unlikely that even a Robbie could move about as he wished through such a well-structured system.

The rumor mill ground true and one morning there was Robbie. A smallish youth with dark collar length wavy hair and regular features, not handsome but clean-cut, the kind of kid most parents would trust to date their daughter. I asked him to take a seat at a back table to which he sauntered casually, a confident cockiness in his typically teen bouncy stride. He chatted with nearby youths, who already seemed well-acquainted with him, without regard for any possible classroom decorum I may have decreed prior to his camp debut. Robbie seemed so at home in my classroom that when I asked him to come join me in the chair by my desk, I very nearly raised my hand to request his permission to speak.

Robbie's exaggeration and boasting was certainly not new to me. I had seen such slickers before who, thanks to the gullibility of our inexperienced youthful population, were able to promote themselves by lying outrageously. One youth who had spent a few weeks in the Navy before being ejected as underage and under-qualified convinced all the boys and even one staff that he had risen to the rank of admiral and was only discharged because the fleet didn't have enough ships and could not therefore assign him one. We had bronco-busters, gun smugglers, auto racers, and authorities on every known subject. I was sure that Robbie would turn out to be like those others. He may have stolen a CB or two, many of our kids then had, but his reputation for being able to "chump" staff and set his own program goals were hard to believe.

I wasn't concerned that he would manipulate me because I knew that I held all the aces. I set his goals and coordinated them with his counselor. I would decide when he had achieved his goals and recommend him for release. Robbie would have to do my program my way. I knew he couldn't beat both me and the system.

"Oh Robert," I said, motioning him to the chair at my desk. "Please sit down here and tell me about your school aspirations."

He took the chair on the right side of my desk facing me and replied, "What do you mean 'aspirations'?"

"I mean what do you want to work toward here at camp, high school graduation or a G.E.D.?"

"Nothing," he said, "I don't intend to work on anything here. I'm just going to do my time."

"Well," I said scanning his transcripts, "you're going to get bored just hanging around doing nothing for eight months."

"The judge said six months, and I already done half that."

"But you won't be eighteen for eight more months and if you don't accomplish anything you'll be here at least that long. I know because I recommend you for parole based on what you've done in school, and if you don't do anything then I won't either."

"I don't like school. I don't like teachers, and most especially I don't like you!" Robbie said.

"What don't you like about teachers?" I inquired growing angry but trying hard to conceal it.

"I don't like the way they teach or what they teach."

"So you think you could do better?"

"Yeah probably I could," Robbie responded sullenly.

"Well, I'm going to give you the chance to do just that," I said taking several paper bound booklets from a desk drawer.

"You see these? They are the Oregon State Department of Education standards." I said, "They tell you what you have to know to graduate. Now you've got three credits to go for graduation, senior studies, English four, and U.S. history." I marked some pages, "Page 37, 39, 40 and in this one page 80 and 81, and here the first four pages. Okay take these and here's paper and pencil and design your own course of study. Tell me what you're willing to do to graduate?" I pointed to the table where he had been.

He looked blankly, then taking the books and papers I had given him to the back table he sat quietly staring at me. I studied him over the top of a book I was supposedly reading. My tactic wasn't new, I'd used it before. Mostly it was just for show. Students could see that to earn credit they were required by some higher authority than me to achieve and, after an hour or two, that they weren't capable of planning their own course of study. I had decided to let him sit there the whole day if necessary for him to admit defeat and yield to my direction.

He picked up one of the books, thumbed through it, and wrote something on his paper. A few minutes later he asked me if he could have copies of the textbooks he would be using. I gave them to him. He scanned them and wrote some more.

That afternoon he laid several sheets of paper on my desk. They were filled with scrawled notes in a crude outline form.

"Well, sit down Robert," I motioned him to the chair at my desk. "I see you've come up with some ideas for what you want to study here. Hummm, the Constitution and organization of the federal government. Yes, that's in second semester U.S. History. Yes," I said turning to the consumer problems, "How to buy a car," I read and inquired, "Would the price include a CB?"

He ignored my comment, but it served to let him know I was on to his reputation and was not impressed.

"And in English I see you want to work on writing and expanding your vocabulary. Humm, not bad, but there are a few important things you left out, like . . ." I thought, "Here's where I lose him."

"Like what?" he said.

"Like the whole Civil War," I said. "You have to at least know something about that! And what about various kinds of financing available to buy that car?"

"Well okay," he replied. "That stuff sounds pretty boring but I guess you're right. Yeah add those."

"And a few other things too," I said. For the next hour we refined his outline and added a quarter credits for typing, P.E., and first aid. He said he'd like to type his work so we figured he could earn a half credit for learning home keys and typing at least twelve words per minute so I gave him a typing instruction book to add to his already sizable collection.

Robbie's work was carefully and precisely done. Several of his course outlines and charts and graphs became a regular part of my instructional materials collection. Robbie was interesting to work with because of his inquiring mind. There were often occasions when he would disagree with something I said, but he did so courteously and several times proved me incorrect. Robbie had become the ideal student, who not only made up his own course of study, but expanded it imaginatively and improved on mine.

Robbie hadn't lost his egocentric desire to be self-directed. He hadn't really given in to the system as long as he was "doing his own thing." After graduation and his eighteenth birthday, he decided to join the Army so he spent his last several weeks practicing ASVAB and various other Army aptitude qualifying exams.

***

Robbie did go into the army and about a year later I heard that he had qualified for officers' training. Someday I'm sure he'll become Chief of Staff—maybe even President—'cause it's got to be done his way.

From left to right: unknown youth inmate, John Forbes (work supervisor), and Denny Meredith (Group Life Counsellor).

CHAPTER TWENTY-ONE

TURKEY DAY

"I'd like to invite a boy from Camp to join us for Thanksgiving dinner," I told my wife Fran. "We have some new kids who won't get home visits. They should be with a family."

"Well, I don't know. The boy we had last Christmas just sat around the house hardly talking to anyone, and then he ruined Joshua's Etch-a-sketch and toy chain saw. Besides you never told me he had killed someone."

"I guess I forgot about that, but I didn't think it mattered what he'd done as long as he didn't do it again," I said.

"I don't think he had much to do all by himself," Fran commented. "Might have been different if he'd had a friend or someone to talk to besides just us."

"Yes, that's true," I agreed, "and there are two boys that don't get to go to their homes for Thanksgiving. We could invite both, but if we do I'll have to watch them closely though because I've found that two delinquent heads can think up trouble twice as fast as one."

Thanksgiving day dawned bright and clear as spring-like as only the rare fall day can be on the Oregon Coast. I drove out to Camp at ten-thirty thinking that would give the two kids a holiday chance to sleep in and since we weren't planning dinner until two or three in the afternoon there was no need to get them sooner.

The two had just finished a late breakfast and were cleaning their dishes while Herb Helstrom, the Assistant Director, sat in the office chair smoking and reading the newspaper, a position he seldom moved from. Herb could run the whole camp from that chair, directing the work details of the various clean-ups, keep order, and even break up fights in a stern voice without even snubbing out his ever-present cigarette or laying down his newspaper.

Herb folded his newspaper subway style which he then could somehow read like one reads the pages of a book, as many times as I watched him I never could duplicate it. I always only went to the same page I started on. Herb was a New Yorker who had come west after working in some of that state's worst juvenile facilities.

When Herb's wife died shortly after coming to Oregon. After that his only interests were smoking and betting. Every racing season he spent much of his time at Portland Meadows Track and his vacations were spent at California's big-time tracks.

One of the big reasons Herb never moved around much was his emphysema. He wheezed and couldn't get his breath, and after his retirement before he died from emphysema, offered himself publicly as a "why not to" smoke example. By then he couldn't speak without gulping oxygen between phrases, but nevertheless, he appeared in the newspaper, on television, and when he could visited schools.

A year or so later Herb was assaulted by three boys in a situation much like the one he was in that morning when I picked up the two boys. On a home visit day with only five boys in Camp, he left his desk and his newspaper to quell a faked disturbance in the dorm. Two older youths, who two days before had been sent to the Camp directly from the Secure Treatment Center for emotionally disturbed youth at Salem, hit him with a two foot long wooden handled garbage can cleaning brush opening an eighty stitch gash in his scalp. That incident ended Herb's lifetime of work in juvenile corrections.

But that morning I greeted him, neither of us knowing all this would come to pass, "Hi Herb, who's left in camp besides the two I'm going to take with me for dinner?"

"Just them two," he responded, looking over the tops of his paper and his half lens reading glasses, "an' after you take 'em I'm going home." He held a cigarette to his mouth as he always did pinched between his thumb and forefinger with his other fingers flared up and out.

"Okay Herb," I said motioning to the two youths who were standing at the counter to follow me.

Both were what we laughingly called "ankle biters" meaning younger, physically small institutional first-timers. They were fourteen-year-olds. The sort of kids the unwary think they can take into their homes as foster children, but which are actually the most difficult inmates in juvenile facilities because they do not yet realize what's ahead. Most have been expelled from school repeatedly, placed on probation several times, and yet are sure that they have outsmarted school officials, police and probation officers, and can also "chump" the institutional staff.

"So if you're so smart what are you doing here," I used to ask them after I had exhausted all my wits, resources, and patience. While probably not the first time they'd been asked, it was the only question for which they didn't have a ready answer.

One of the boys was African-American and the other an institutional "seen a thousand of 'em" kind of kid. Neither one had any unusual offenses (I checked so I'd know about allergies, deviancies, and home towns) to know how much freedom to allow and where to send the cops should one or both runaway. They had been truants, petty thieves, and assaultive, rebels to authority. These kinds of kids are generally either neglected or pampered. My two youths were obviously neglected and had probably never eaten Thanksgiving dinner in their own homes much less as invited guests.

The twenty minute drive to my house was quiet as the boys talked together about mutual acquaintances they had made during their stay in reception cottage. I didn't attempt to enter into their conversation because it was the same as many I had heard before and listening often times helped me gain some insights into students' characters. However, most of what I heard was quite disappointing news so far as their rehabilitation was concerned.

After we arrived home I introduced the two to our family members: my wife, her eighty-eight year old mother, our five-year-old son Joshua, my parents in their seventies, our two college age children, two visiting nieces and a nephew, also college students. Certainly none of these had much in common with our two visiting youths so I suggested that they might go play basketball.

I gave them a basketball and took them to the hoops which were about a city block from our house and at a lower level so that I could see some of the paved court and could see the ball when shot at the free-standing back boards. I told them to come back to the house every thirty minutes and check with me to see whether dinner was ready yet.

"What about the wine?" Fran asked.

"What about it?" I retorted, not knowing what she meant.

"Shall I serve them wine with their dinner?"

"Sure, just the same as Joshua, a small glass cut with some water. That's okay," I answered.

Fran was then making wines. The one we were having that day served in small long stemmed wine glasses was made from rose hips and was a beautiful crimson color. It was delicious and certainly low in alcoholic content.

Fran always liked to set a nice table with her best silver and glassware. We used some table service during these holidays which as in most homes sits idle in cupboards the rest of the year. This was the principal reason I had invited the Camp boys. I thought at least they could experience a family holiday dinner instead of eating their usual camp fare which although wholesome was drab.

"The boys are down at the school grounds playing basketball, but I told them to check in every half-hour so whenever dinner is ready so are they," I told Fran.

"Every half-hour," one of the nieces said. "You aren't taking any chances are you?"

"Well, better safe than sorry," I responded. "You know in Garibaldi that's long enough to go from one end of the town to the other and back again. That may even be too long. Maybe I should have said fifteen minutes. Anyway I can see the back boards from the living room. When balls stop bouncing off them, I'll go down and check on those two."

Dinner was a little later than I had thought, but Alvin and Tyrone kept shooting and checked in like clockwork every half-hour. We had a nice meal and I had the boys back in camp by four. They asked to go back saying that they wanted some time to play pool and write letters.

Sunday evening Cecil Oman called, "Jack do you know anything about this stuff I've taken from Alvin and Tyrone?"

"No," I replied. "What kind of stuff is it?"

"Mostly old coins, you know like foreign—some is Chinese or something, the kind with the holes in them. And the jewelry is mostly old and cheap, costume jewelry I'd say. They said a friend of yours gave it to them."

"They didn't see anybody or go any place that I know of. I let them go down and play basketball, but except for that they only saw my family," I replied.

"Okay thanks, I'll check it out," Cecil said.

Both were returned to detention at MacLaren that evening.

The next morning when I went into work, Dick White said, "Did you know that you had a crime wave in Garibaldi last week?"

"None that I know of," I replied. "What kind of crime wave?"

"The Old House, the Old Barn, and the Train gift shops were all ripped off."

"Oh yeah, by whom?" I asked, even though I had guessed "by whom."

"Tyrone and Alvin were busy boys."

"Couldn't be them," I argued disbelieving. "I didn't let them go anywhere except to play basketball within sight of my house."

"Well they met another kid from Mac' who was on home visit and together with him they broke into three shops and maybe some houses too. I hope you're pretty good friends with your neighbors?"

"I used to be. Maybe not anymore though," I said. "I guess I better go find what they stole from whom before I answer that."

I called the owner of one the three shops that were broke into only to discover that one person owned all three. Apparently these were the only places Tyrone and Alvin had burglarized, and the owner said that they had entered in one case by removing a broken casement which had only been visible as being defective from the inside, in another they had come through an attic opening, and the third was a half-mile out of town and not very noticeable as a likely spot to break and enter. This was all very suspicious considering Alvin and Tyrone had checked in with me every thirty minutes while they were supposedly playing basketball.

"Well Jack I think we've solved our mystery," Dick White said a couple of days later. "Tyrone and Alvin told the Adjustment Committee that they just happened to meet one of the local kids who was on Thanksgiving home visit from Mac', and he told them about those places they burglarized."

"Told them nothing, he had to be with them. In fact I don't even think our two boys were in the Old Barn. They just didn't have time to go that far and get back to report to me," I replied.

"Well maybe," he said, "But did you know you're being sued?"

"For what and by whom?"

"Contributing—giving a minor alcohol. By Alvin's mother who blames you because he told the Committee you got him drunk."

"He had four or five ounces of watered wine . . . and after the burglaries. My five-year-old son drank the same amount," I responded.

"Well maybe, but his parole officer is a teetotaler, and Legal Aid says they will handle the suit for the mother. With his P.O. backing them it could mean trouble for you," he commented. "We'll just have to wait and see."

***

Several months passed. Nothing came of the supposed suit. Almost a year later I inquired about it and was told that it had been dropped because the legal aid lawyers could never find Alvin's mother at home. She was always at the tavern. Alvin kicked around through the system for years and may still be. That was the last time I invited inmates to my home.

CHAPTER TWENTY-TWO

DEAR QUEEN:

"Mr. Graves, how do you address the queen?"

"What queen?"

"Elizabeth the Second, Queen of England."

". . . and Scotland and Wales," I added.

"No, just the Queen of England."

"I don't write her so I've never addressed her, but I have a book which will tell us how. If you really need to know I'll get it for you."

"Okay, yes, would you please?"

"Why are you writing the queen?"

"To help me get out of here."

"She doesn't have any say-so on this side of the Atlantic and not too much even on her side so it won't help."

"She could write President Reagan, and he could pardon me."

"Well, she's got so many problems with her own kids that she's not going to take on any more, and too, Reagan can only pardon federal prisoners, not state."

"Oh."

I got the book I had told him about and we looked up the correct form of address, "Your (or Her) Royal Majesty, Queen Elizabeth the Second, Buckingham Palace . . ."

"The Governor could pardon you, but I don't think he will."

"Well, I'm still going to send this letter. It can't hurt!"

"Might get you a ticket from here to the 'funny farm', but if you get a reply I'd like to read it."

"Okay, what did you say the governor's name is?"

"Victor Ateyeh."

"What's his address?"

"Salem, Oregon."

"Zip code?"

"97301."

"Nathan, I can't prevent you from writing the Governor or anyone else you want," I said recalling *the lawsuit* in the late seventies filed by Legal Aid lawyers on behalf of four youths listing over ninety complaints against Juvenile Corrections policies and procedures. The Attorney General's office gave us a list of guidelines to follow to avoid further violations of children's rights. These prevented us from reading or halting our kids' incoming or outgoing mail. We even had a youth stalker whose sexually explicit and threatening letters to a little girl had to be mailed (Ironically when the Oregon Supreme Court finally ruled on *The Lawsuit* in the early 1990's twelve years later, all but a half dozen of the Attorney Generals's directives were thrown out).

So Nathan wrote the Queen and the Governor, and oddly enough even the Duke of Windsor. He wrote people all over the world asking them to help him regain his freedom. It didn't help him get out, but one youth who was a stamp collector added considerably to his portfolio. He did show me his answer from Buckingham Palace which was, to his disappointment, signed by a personal secretary to the queen. As expected, she thanked him very much for writing but said she had no authority to free him.

"Jack are you the one encouraging Nathan to write all these 'screwy letters'?" Larry Morgan, our counselor, asked. "Nathan says he's writing them for school."

"Well not exactly," I said.

But before I could reply he added, "The Governor's Office has complained to the Superintendent."

"I can't stop him, and he's supposed to be studying for his G.E.D. so I figured why not show him how to write letters correctly and so use them to work on sentence structure and composition. Writing the Queen is really quite harmless."

"So you are helping him write those 'nutty' letters!" Morgan exclaimed.

"Sort of," I conceded. "But you know he's reluctant to study for his G.E.D. and this is a way to help him prepare."

"Well, find another way, will you? You're right, we can't censor his letters or prevent him from mailing them, but if he has to write them on his own time, maybe he'll quit writing them."

"Okay, but I don't think it's going to even slow him down."

I was right. He wrote just as many letters as before and did less class work. Nathan didn't want a G.E.D. He contended that the Certificate was useless since he would only be tilling the soil behind a horse or mule. He had come to Oregon from Pennsylvania and had become a street kid. Nathan was nearly nineteen and had so few credits that finishing high school was not a viable alternative.

We used every kind of inducement, alternating threats with pleas to get him through four of his tests. He left camp only half-educated without his G.E.D. of which he had passed everything but science. However, he had become an excellent letter writer which I hoped would someday prove helpful.

***

Several years ago Nathan was interviewed in the business section of the *Oregonian* as the new manager of an adult arcade business. I assume that he applied for the position by letter.

CHAPTER TWENTY-THREE

T-MAN

Tyson was older and seemed more mature than most youths who came to camp. In fact on his first day at camp he so impressed Mr. Hogland with his diligence and cooperativeness that Doug, who generally adopted a skeptical "show me" attitude, awarded him top conduct grades and lavish praise for his excellent work at the Y.M.C.A., where we were cleaning the handball courts as a part of our membership fee waiver agreement.

Tyson was very reserved, but not so much so that afternoon during my pre-placement interview he confided his ambitions to me. It seems that his girl friend recently bore his child. Tyson said that he wanted to get his G.E.D. and then go on to college and learn a business or a trade. He wanted his child to grow up with a real father who came home after work to his supper and slippers. This aspiration was an uncommon one for a black youth, but I encouraged him and urged him not to give up before he made it.

The other black youths were quite deferential to Tyson. I attributed it to physical dominance since he was larger than they. He reminded me of a young Malcomb X, tall and slender, but well muscled, with wavy hair that followed the contour of his head, and he wore rimless spectacles like Malcomb X had. It may well be that Tyson deliberately cultivated this image for as I was to find out soon enough he too was a special kind of leader.

For several days he made great progress and it seemed likely that he would be ready to take his G.E.D. test at the time he and I had agreed, but then he quit studying completely. I gave him lower performance grades which caused him to become so angry and so defensive that he did even worse.

During "the grading period", the last fifteen minutes of each day, all boys who had been in school that day assembled in a classroom so that we teachers could ask each how the day had gone for him and give each an opportunity to report what he had accomplished that day. It was accountability time for those who hadn't accomplished much. Some would try to bluff their ways through imaginary accounts of great achievements, but most knowing the "jig was up" were forced to report that they had done little or nothing.

It was also a time for anger and hostility because this too was when we teachers "graded" our students for the day—one grade for achievement and one for attitude. Very few of our inmates were competent liars regarding their days's work. When "put on the spot" some of the more indolent ones would resort to making excuses and when that failed, verbal aggressiveness. While I originally began this practice to stimulate performance, it also proved useful as a test to discover those who lacked restraint and self-control and so was very helpful in assessing our inmates' abilities to meet frustration and stressful situations after being released.

Tyson's first day's grades were excellent. But then when they began falling, he grew more sullen. He didn't complain at first or argue his case as he was entitled to, but at the start of his third week in school, he erupted! This was the beginning of the end for Tyson.

It was the end of Tyson's school performance as well as the end of his good peer relationships. It soon became apparent that he was physically and mentally dominating the younger and smaller black youths. It was as if he had entered into a severe state of depression and needed to infect everyone around him. I asked him to accompany me to Mr. Morgan's office on two or three occasions where the counselor and I urged that he return to his former successes so that he could be released to return to his girl friend and their child as he once had said he wanted to do.

"Man, Mr. Graves! You betta quit standin' on T-Man or he's gonna bust on you," one of the younger smaller black youths said.

"T-Man is Tyson I guess," I replied.

"He sure is. Ain't you seen T-Man writ' on the walls in the west tunnel an' on buildin's all over town?"

I remembered that I had seen that scrawled often in dribbly, runny spray-painted, crude lettering. "Yes, I've seen it," I said.

"Well that is Tyson. He is boss man of the Crips."

"There is the Crips and the Bloods. Right?" I asked, well aware of the answer. "Any other gangs?"

"None that nobody would admit to bein' if they wants to keep livin'," my informant replied. "Tyson he is T-Man, *The Man*, you know what I mean, an' his place is bein' took over by a dude up from Compton. When Tyson gets outta here, and it betta' be soon, he gonna take over from that dude who jes' may nevar be seen aroun' Portland town no moe. Least ways he take 'em out with a gang-bang. But you is keep'n the T-Man locked up so's this Compton dude can take over and makin' T-man awesome pissed off at you."

"So if I lay off of Tyson and let him do what he wants, how does that help him get back on the street?"

"He wants you to send him to pass his G.E.D. test," the youth responded.

"Well, I could send him to take his test since he's eighteen, but for sure he won't pass unless he gets back into studying like he did when he first came here. So what happens if he flunks the test?" I asked.

"He runaway, that's for sure, guaranteed!"

"Why are you telling me this?" I asked. "T-Man sent you didn't he?"

When he failed to answer me I knew I was right.

"Well tell him that I won't let him take his G.E.D. test until I'm sure he's going to pass it and that's only after he's studied like he agreed to, and if somehow he scams his way into testing and flunks, I'll make sure he studies six months before he tries the test again! Tyson may be T-Man to you and on the street, but he's no big man to me! Just another student in my class."

After he left, I mulled what few options I had; I could allow Tyson to test, but if I was wrong and he passed there would probably be another killing in Northwest Portland, and if I was right and he didn't pass, he'd runaway with the same result. He was probably safer in MacLaren's gangs unit so I reasoned it seemed best to recommend that he be returned to MacLaren before he could runaway to kill or be killed.

The counselor and Camp Director were both gone but I resolved when they returned to recommend that Tyson be returned to Woodburn and placed in the new gangs unit, but before I could an incident occurred which made my plan impossible.

At that day's "grading period" when he was to report what he had accomplished, Tyson erupted. I think that my comment about him flunking his test and remaining at Camp even longer had been reported to him. As he passed my desk leaving my classroom after the grading period he stopped and glared at me. His fists were clenched and his body coiled ready to spring.

He snarled, "I may kill you, Mr. Graves! Someday, maybe today or maybe tomorrow, but when I do, it will be because I can't help myself. If I lose control, it'll be your fault if I do, 'cause you're pushin' me."

He was trembling, shaking with anger and frustration.

"Tyson if you kill anybody—me or anybody else—it will be because you want to, because you've consciously decided to do it! You've got perfect control and you know just what you're doing and why you're doing it. So if you do, you'll pay the price for it. Life in prison maybe, or even the gas chamber. So if you're going to kill me, do it now, or shut-up and get your ass out of here!"

I called his bluff, not really sure that he was bluffing, but aware that it was a good time to do it. There were other staff in the building who if needed could come to my aid, and there were ten or a dozen youths, several of them probably gangsters, who needed to see the T-Man cut to size!

***

That night he ran away with another gangster. They were quickly picked up and were returned to the gang unit at MacLaren. T-Man apparently lost control of his gangsters for the next I heard of him was in the newspaper. Three youths had attempted to rob a mom and pop store. The three beat the elderly owner badly, but the owner's wife who was in the back brought a gun out with her. She shot one youth killing him, and wounded another who bolted out the door. Tyson was found huddled in a corner pleading for his life when the police arrived.

Camp Tillamook as seen from the service road.
Hanger A is in the background.

CHAPTER TWENTY-FOUR

THE AMERICAN WAY

David Sung's family came from Taipei, Taiwan, when David was in primary school. His father had always wanted to immigrate to the United States and saved as much as he could from his meager factory worker's earnings to make his dream come true. Mr. Sung took night classes in English hoping that he might someday find a job sponsor in the U.S. David's mother was fearful. She was frightened of moving to a strange land with strange customs. She would have preferred to remain in Taiwan, but found it unthinkable to deny her husband's dream.

Life in America was not Mr. Sung's utopia, but he did find work on an electronics assembly line, not so much different from what he had done in Taiwan. His wife worked part time as a motel maid changing linens and vacuuming. David was set back from the third grade to the second grade, but he learned the language and wanted even more than his parents to become a "real" American. Instead David grew more ashamed of his parents. They didn't have the time, and David discouraged them, to go to school conferences so they could learn about David's new friends.

In ninth grade, crisis faced this little family of new Americans. David was arrested and placed on probation as a minor in possession of drugs as well as for drinking on school premises. His mother was bewildered about what it all meant and she counselled return to Taipei, but David's father knew how American parents dealt with such problem children. A spanking was what David deserved and so Mr. Sung vowed that he should have one the very next time he disobeyed.

That happened too soon. Mr. Sung took off his belt and informed his son that he was going to be disciplined American style. David resisted but Mr. Sung prevailed, and in what became a "donny brook", left bruises and a mouse on his son's cheek. David's probation officer had more up-to-date information about the American way than did Mr. Sung, so she filed charges against him for abuse little knowing that David would be removed from his parents' home. "Temporarily", the court order had said. So David was "temporarily" placed in a group home for drug addicted youths. Thus began the Sung's long struggle to become reunited as a family. David asked to return home, but at that very time was discovered to have used drugs in violation of his probation.

The Asian gang with which he affiliated wore long oversized overcoats with their collars turned up to the brims of felt hats. Their overcoats were more than just their "flag" of uniqueness but served as well to hide weapons and stolen goods. David wasn't allowed to return home, but his gang life replaced his family life as he moved from home to home always meeting another "Fagan" to teach him better ways to steal and drink and use L.S.D. and cocaine. Seven months after he was taken from his home he was in Juvenile Detention Hall and three months later two more foster homes, then MacLaren, and finally two years later he came to Camp Tillamook garbed in his overcoat and felt hat, under which he oddly wore a blue navy knit cap, he said because it was winter.

"David Sung came to class wearing an overcoat and hat. I don't permit hats in my classroom and there's no reason for him to wear such a heavy coat since it's quite warm inside," I told Mr. White, Camp Director, on David's first day.

"They're his personal identification symbols. Let him wear them," he said. "He's in a foreign land and he needs something to hang on to here. Have you read his family's story?"

"No," I said, "but I think his coat and hat are gang symbols. We don't permit anyone else to wear gang flags. Why him? You can't convince me that's native dress in Taiwan."

"He's isolated from the other boys here and lonely. Make an exception. Let him wear his overcoat and hat. What does it matter to you what he wears? Besides there's the Attorney General's rulings about student's clothing rights . . ."

"He's isolating himself further, hiding under that coat and hat, and well . . . I'm not even sure what he looks like."

"A day or two more when he's better acquainted he'll take that stuff off, you wait and see."

The next Monday when he reported to work crew in his strange garb, he was told he couldn't go on to work unless he wore the standard blue work shirt, jeans, yellow slicker, and sou'wester hat. He refused and so didn't go to work that week. In school he accomplished little. I informed him that since he was not working and not studying in class, I would recommend he be returned to MacLaren. It was an empty threat for that's what he wanted, return to a non-school program where he didn't have to work and could wear his big overcoat and felt hat—if only to watch television all day.

Then one morning he didn't report to school. I found him in the dorm . . . in bed. "What are you doing in bed? Are you sick?" I asked a shapeless mound in a top bunk which I assumed to be David.

A muffled voice replied, "I'm not coming to school today."

"Why?" I inquired.

"Tony took my coat when he ran last night."

"Your overcoat?" I asked.

"Yes," the lump in bed began to sob. "I have to have my coat. I haven't gone anywhere without it for over a year."

"Well, it's time to start. I read your history, how you charged your father with abuse because he hit you, and that your parents have been trying for two years to get you back. Do you want to go back to them?" I asked.

"Yes," he said, pulling back the covers a bit so he could see me. "I've told everyone I could I want to go home to live with my mother and father."

"Well, not as long as you wear that overcoat and hat," I said. "Nobody's going to listen to you because they won't believe you can get along with your father. Now's the time to prove them wrong. Come to school and go to work crew and do the program here like you're supposed to—then you can go home from here."

"I'll try," he said.

And he really did, for a time. David did very well in school. He was remarkably intelligent and was gaining ground on his lost years of education. Then, two months later his overcoat was returned to him after Tony was recaptured. David ran away that same night wearing his overcoat.

***

The last time I saw him was in a photograph. Doug and I were attending a Portland Association of Teachers workshop on "Gangs in Portland." A police officer from the gang unit had a table spread with photographs. There was one of a slightly built Asian clad in nothing but skivvies. On a white sheet before him was displayed a collection of weapons, firearms, knives, even Ninja throwing stars. David was charged with fourteen serious offenses, including driving without a license, DUII, eluding a police officer, extortion, assault, and auto theft. There was no overcoat in the picture.

CHAPTER TWENTY-FIVE

A MOTHER'S ANGUISH

Jimmy Williams was impulsive. Everything he did was momentary. His mind darted from one negative notion and on to another. Jimmy could be funny, not intentionally, but his "off-the-wall sillyisms", although amusing, often put us at odds because he disrupted the sacred sanctity of study. Then to make matters worse just when the class had calmed, he would innocuously inquire, "Hey dudes, who'ya laughin' at?" re-triggering the disruptive mirth anew. "Immature" was the key word used in his monthly progress reports, "Jimmy continues to exhibit immature levels of thinking resulting in poor decision-making so he is often involved in mischief and misconduct," and I added month after month, "For Jimmy it's now or never. If Jimmy is not to become an habitual offender." As I was to discover, it was already far too late.

Physically he was a butter ball. Short and stocky, he wasn't just chubby, for Jimmy hadn't shed his layer of baby fat. A lot of fourteen-year-olds are like that. Everyone can see they're on the verge of manhood yet their dimpled smooth cheeks deny their entry into masculinity. Jimmy's smile was most engaging because his face was perfectly round and when he smiled he looked like the smiley face cartoon one so often sees. It was hard to believe that this funny little guy was already a hard-core gangster.

Jimmy's delinquency history was extensive but minor, several shop liftings, a lot of truancies, and even more curfew violations. He had been committed for dealing drugs. Jimmy was learning the drug business from experts who came from L.A. bringing their crude gangster mentality with them. Crack cocaine had recently hit the streets doing for dope what Henry's "T" did for autos. Affordability made every street corner and playground a marketplace. Unlike heroin, coke, and a variety "high society drugs", crack was cheap and versatile, a drug which could be used in many forms and was available to anyone with five bucks.

Jimmy had no male role model. His mother was raising him alone, and at the same time, trying to earn a living for both of them. She feared for his life. She was sure that he was going to be killed in the gang strife which had snuffed out so many young lives. Jimmy was a "wannabe" which means on his way into the gang life. His mother had sent him to relatives to live, but not far enough. He was back in town the same night. She was desperate to protect her son, even attempting to prevent his release from MacLaren, but she was only able to stall for a few weeks.

"Hey, Mistur Jack," he started with a typical Jimmy-ism, a mixture of formality and familiarity. I didn't deny the use of my first name to students, but over the years they had come to address me as "Mr. Graves", and unlike many other staff I believed their formality made it easier for me to maintain a learning structure in my classroom.

"Yes, Jimmy?" I replied.

"Did you know George Washington had slaves?" for once he was asking me a question which related to his eighth grade history.

"Yes, I knew that. He lived over two hundred years ago when slavery was legal. A lot of the founding fathers were slave holders."

"Was Thomas Jefferson?" he asked.

"Yes him too," I replied thinking at last we're going to get some positive education going. Jimmy is actually talking about what he has just read in his history book.

"I got a friend named Washington, an' Jefferson, you know he's a rich black guy on tv. Maybe they had ancestors who was slaves on those old dudes' plantations," he observed.

"Could be Jimmy, a lot of the slaves took plantation owners' names, because when they were freed, many didn't have family names."

"Yeah, like in *Roots* when they made the black guy let 'em call him 'Toby' instead of 'Kunta."

"Yes, like that, but after generations many African-Americans forgot their African names, and when they were freed, many wanted to be like the white people and have real families and . . ." Jimmy interrupted me. I knew I had over-reached Jimmy's need for self-expression with my lengthy response.

"I don't want no real family . . ." Jimmy commented, looking expectantly to me for rebuttal to his shocking comment.

This was an old avoidance stratagem Jimmy had found successful before, the intellectual equivalent of "so since this is starting to bore me, in your face man!"

"Slave holders frequently tried to prevent slaves from having family relationships so they cou . . ."

"You got a family, Mr. Graves?"

"Yes, a wife and three children."

"How old are your kids?"

"Two grown and one in high school," I replied. "But let's get back to history. It's a lot more interesting than my family."

"You have any other ladies besides your wife?"

"No," I replied, "just her."

"Must be borin' havin' sex with the same woman all'a time."

"She's my best friend too . . ."

"Man, your wife is your friend!" he exclaimed. "I got all the friends I need, and when I get to be THE MAN, I'm gonna have all the ladies I want whenever I want 'em . . ."

This time I cut him off, since our conversation had side-tracked on negativism, "But don't you know Dr. King and a lot of others fought so you could have a wife and children, and a family name, and . . ."

"Hey man, look what happened to Martin an' most them other 'good-goodies'—dead ain't they? When I die it's gonna be for me gettin' rich. I ain't never had no father, and so what? I know how to get by on my own."

"Well, you've got a mother, and she's worried about you. I know because I've talked to her on the phone. She's afraid you're going to get killed in a drive-by shooting. She loves you and she doesn't want that to happen to you."

"You hear of Marsha Johnson?" he said.

"No, who's that?" I asked.

"My couz who was shot in the back of the head last year and dumped in an empty lot on Columbia Boulevard."

"Who shot her, do you know?" I asked.

"She was workin' for the big man of the Comptons, and then when she stole his money for drugs 'cause she had a heavy habit, he said to me, 'Jimmy she's your couz so here's a chance for you to get big with us', and he gave me the gun."

"And you shot her?"

"I ain't saying, but me an' two friends dumped her body. That's why my mother is 'fraid."

"How old was your cousin?"

"Seventeen," he replied.

"I think it's time for you to get back to your studying. If you want to talk about Washington and Jefferson or slavery, okay, but no more of this 'gang' talk. It's dirty and depressing."

"You come to big town in five years and I'll get you all the ladies you need, and since we're friends, maybe even for free, okay Mr. Graves?" Jimmy's face glowed with one of his happy face smiles, somehow now sinister.

"I think your mother is right Jimmy. In five years you're going to be dead. Probably killed in a gang-bang, and if not that, you'll be in prison."

"My mother, she don't know about my couz. You don' tell her, okay?"

"No, I don't believe you anyway. You're just a little boy to me, a 'wannabe' with a big mouth."

***

Jimmy's parole counselor verified that he had probably murdered his cousin the year before. He and two other youths had been seen talking to her the night she was killed. Although Jimmy was questioned regarding her death he never admitted to knowing anything about it.

Less than five years after our conversation, I read in the newspaper that Jimmy had been killed. It happened in a drive-by, a "gang-bang", as his mother had feared. Jimmy had apparently achieved the success he wanted in the gangs for the paper reported him to be "an important gang leader."

CHAPTER TWENTY-SIX

BUCKY HENSON'S AILMENT

"Henson's in the dormitory. Off work today as a 'sicky'." Denny Meredith told me when I came in.

"I hope he's really ailing, because I got better things to do today than run back and forth between my classroom and the dormitory," I replied.

"He doesn't look that sick, but he's complaining of having a sore stomach again," Denny responded.

"Again? He's had several of those. I'll go down and talk to him. There must be something causing it . . . or it's gotten him off work crew before so he's using the same symptoms again," I said taking a fever thermometer from the desk drawer. "If it's chronic appendicitis, it'll hurt low in his abdomen on his left side but much less if he lies on his side and pulls his knees up to his chest. Where does he say it hurts him?"

"I asked," Denny said, "but he ain't much of a talker."

"Knowing you, he probably never had a chance to get in a word, Denny."

I left the office, stopping briefly at the counter to fend a "questioning complaint" from a youth who was dissatisfied with his previous day's grades I had put in the grade book.

"Well then Tony, tell me why you spent most of the afternoon writing a letter instead of studying for your G.E.D."

The curly haired youth replied, "I had'a write my girl friend. It was important, 'cause she's comin' up next week with my parents and I wanted to tell her to bring my guitar."

"Took you all afternoon to tell her to bring your guitar?" I questioned.

"I had to tell her how much I missed her too."

"Well, if you don't get to studying and get your G.E.D. completed she'll be missing you for a lot longer. And no. I won't change your grade for yesterday, but every day's a new day so today's the day to get your grades up—maybe even 9-9's."

I continued on into the dorm looking for an occupied bed among the thirty there. There was someone in a top bunk at the far end of the dorm.

I inquired, "Henson, are you in there?" There was no response from the mound of woolen blankets. I shook the mound gripping it about where a shoulder should be. "Henson, Henson, it's Mr. Graves. I want to take your temperature."

"Uh, okay," the rumpled heap groaned.

"Put this under your tongue," I said handing him the thermometer. "Do you feel feverish?"

"Whacha' mean?"

"Are you warm and sweating?" I asked putting my hand on his forehead.

"Naw, jes' my stomach hurts."

"Where, down low on the left side or higher up?"

"Down low on both sides but more on the right than the left."

"Lay on your side and pull your legs up to your chest."

"Hey man, I'm sick. This is stupid," Henson commented, but he did as I asked.

"Does it hurt as much?"

"Yeah, that don't help none."

"Press in where it hurts."

"Okay."

"Does that hurt more? Is your abdomen tense?" I asked.

"What do you mean tense?"

"Tight like it's hard to press in or hurts when you do."

"Naa," he said, "It feels like it always does."

In my previous twenty-five years I had encountered similar problems and had learned to diagnose kid's ailments more from how they responded to my questions than what they actually said. Several boys had had appendicitis attacks—three times already ruptured. Frequently the problem was indigestion, sometimes from eating bad stuff, like psychedelic mushrooms or chewing up and swallowing half a baggy of marijuana. Then we had two cases of ulcers, fairly rare in juveniles, but both previously diagnosed, and two intestinal obstructions. There were several instances of injuries caused by fists and feet resulting in possibly damaged organs such as livers and spleens. Since most youths were simply malingerers evading work crew by faking symptoms, diagnoses were very difficult, but so far no one had died or suffered permanent disability, probably more a matter of luck than good judgement. I always looked for real ailments when I talked to "sickys" rather than assuming they didn't want to go to work that day. I recalled one instance many years before when I had been ordered not to allow a chronic malingerer to go bed or take him to the doctor, but in violation of those directives, I took him anyway, and by noon he was on the operating table with a ruptured appendix. No one criticized my disobedience.

Bucky Henson was a gangly youth of seventeen with a pimply face and stringy hair hanging across his forehead which while not cut long dribbled wispings over his collar. He was nearly six feet tall with a reputation for not backing down from a fight. He was morose and not much fun to be around so he had few friends. I tried to help him study for his G.E.D.; however, he had resisted my offers saying that he would rather do it, "On my own." Bucky Henson was pretty much a loner—a stranger to me and most in camp.

I didn't really know much about his family or home life for I hadn't read his rap sheets or psychological reports. I only read them when it was necessary because I was a teacher, "I'm just the same as a teacher anywhere else," I told them. "This is our school. When you come in that door, you are a student just the same as you would be back home if you were there. I don't care what you did 'on the outs' because that doesn't have anything to do with school. In this classroom we are not delinquents; we are students."

Students like Bucky were frequently only children of older parents or the last of several children born to older permissive parents. His socialization and sharing skills were primitive "grab and go's." He seemed to me as selfish and ego-centered. Maybe the kind of kid who never had to share . . . or maybe never had anything worth sharing. It was time to find out.

"Did you take any aspirin?" I inquired.

"Naw."

"Well it might help ease your pain, but take 'No Aspirin' instead of aspirin."

"How can I take no aspirin to help my pain. That don't make no sense."

"'No Aspirin' is a brand name of an aspirin substitute. It may be safer to take than aspirin since we don't know what's causing your pain. I'll send you down a couple and a glass of water, okay?" I said as I left.

At the counter I put the two pills in an envelope and asked a youth nearby if he would take them along with a glass of water to Bucky Henson in the Dorm and pushed the buzzer to start school classes. Then I had an idea, I wanted to learn how many times Bucky had used these same symptoms to get off work crew so checked the previous work rosters and discovered that he had complained of the same kind of abdominal pain three times before—each time almost exactly a month apart.

At morning break I checked Bucky who was sleeping and stopped at counsellor Larry Morgan's office on my way back. I found him smoking his usual cigarette and reading the morning newspaper.

"I see you're busy as always," I commented, searching for some "nastyism" appropriate to our usual repartee.

"Yes," he said. "I'm reading about the Portland teachers' latest demand. It's something you might ask for."

"Which is?" I asked, stalling, knowing that some insult would follow his seemingly innocuous comment.

"Cops in the hallways. They want cops to keep order in the hallways. Whatever happened to hall passes?"

"So now students show their passes to cops instead of hall monitors so teachers can sit in offices like counsellors smoking and reading newspapers. I like it. I think I'll propose that we get a cop Rick Hill or Bob Jester. I've been working too hard."

"So if you're so hard working what are you hanging around here 'bugging' me for? You've made me lose my place in this article." Larry nagged.

"Bucky Henson's in bed again," I said.

"So . . .?"

"He's taken '24's' the same days for the last three months. He gets sick every twenty-seven or twenty-eight days."

"He's probably just having his period?"

"Be serious, you're the counsellor here. You're supposed to know why kids do what they do."

"Hey, I'm only a counsellor, not God. Go ask the Camp Director."

"What kind of scam do you suppose he's pulling on us? I'm going to take him to the doctor."

"If he's 'scamming' us, why take him to see a doctor?"

"Maybe the doctor will paint his throat or pump his stomach or do something nasty enough to get him back on work crew," I said.

"Graves, every time you take someone to the doctor it costs the state thousands. Take it easy on the taxpayers. Just let Bucky stay in bed a day or two and he'll be all right."

"If you were a real counsellor, like Ray Cullen was, you'd give him a FIRO B or chart his behaviors according the phases of the moon. Now Ray was a real counsellor who knew kids."

"Yes, I might stir up a potion or two of 'puppy dog tails' too. You took the 'FIRO B' too Graves. I've still got it in the file over there. Your 'moon phase' chart shows that you grow hair on the palms of your hands every full moon."

"Well, enough of these pleasantries, I'm taking the kid to the doctor and I want you to back me on it."

"Well all right, but keep the costs down this time. No ruptured appendixes, smashed spleens, or hepatitis cases, okay?"

The examining room was small. Henson and I waited almost half an hour before the doctor came in. I had learned to stay in the examining room with the boy, remaining as the doctor examined the youth. I wanted to be sure that our patients didn't ransack the doctor's medical cabinets looking for drugs or drink rubbing alcohol, but perhaps the best reason was to get it straight as to what ailed our patients directly from the physician.

After checking his pulse and blood pressure and tapping on his chest, the doctor asked him to pull down his trousers for the hernia "cough test". Bucky's back was toward me to maintain his privacy. The doctor and I were eye-to-eye over his shoulder.

"Now cough hard," he told Henson.

"You must know your testicles have not yet descended?" the doctor asked.

"What's that mean?" Henson asked.

"Well we're going to order further tests, but I'm sure that you have two sets of plumbing. Inside you're plumbed like a female and you're having these pains because your body wants to menstruate, but can't since there's no opening to enable that."

"What's all that mean to me?" Henson asked.

"It means you can be male or female. You have to choose which based on your feelings and which life style you think you would be most comfortable as. Surgically it would be easier for you to be a female, but since it's such a serious decision you need to think about it. It's important that the sex change be made soon for your health's sake. The surgery will be easier the sooner it's done."

"I'm a girl?" Bucky gasped.

"If you want to be," the doctor replied.

***

Larry was just leaving his office as we returned. "Well, Graves, how much will this one cost the tax-payer?"

"You were right Larry," I said, "and you didn't even know it."

***

Last I knew Bucky's surgery had been postponed indefinitely. CSD and welfare (this was before the Oregon Health Plan) could not agree about which should pay for his sex-change based on their bureaucratic definitions of essential or elective surgery.

These kinds of situations have been further complicated by the Downsizing Act of 1985 and other subsequent legislation which purports to be a return to local and county government control (and some fear financing) of corrections both at the juvenile and adult levels. Downsizing limited the populations of MacLaren-Hillcrest to 330, later increased 350 (about the same as it was when I started at camp in 1957). As it became apparent that these numbers were too small, special funding (to exceed "cap") was allowed for sex offenders and drug-alcohol abusers. Thus juvenile corrections managers felt compelled to identify as many youths as possible in those categories to make room for the others. These special classes of offenders are held in "treatment" eleven months and longer. There is no such provisions regarding criminal offenders (especially those charged with property crimes) so auto thieves, burglars, and robbers are held shorter and shorter lengths of time as their numbers increase to remain within legal capacity (recent changes in availability of cell space and sentencing guidelines have changed that for adults, but to a lesser extent for juveniles so the legislature, in their wisdom, has simply lowered the age definitions of adult offenders).

As law-makers assume greater decision-making roles State managers are less able to exercise initiative in coping with onrushing legal and legislative mandates so the corrections systems stagnate in directionless limbo between legislative sessions.

Late August 1989, retirement group at Cedar Bay Restaurant. Left to right: Jack L. Graves (retiree), Doug Hogland (teacher), Jim Richard (principal), and Richard L. White (director).

THE SEX-OFFENDER ERA

Our society is obsessed by sex. Pornography and nude dancing are said to be protected by First Amendment rights, but sexual license is not. It is this latter fact which often misleads those young people who commit criminal acts. Today's sex-offenders may range from the curious seven-year-old to adult molesters, rapists, and even psychopathic serial killers.

SO's tend to be lost inside of themselves. Most are "alone in the crowd" in the sense that they live in a dream life in which their fantasies may become indistinguishable from reality. The worst cases are swept into the abyss of maniacal sadism. Why? Nobody knows. It seems impossible to identify sex-offenders until they offend for they run the gamut of physiological and social classes. Then once they have offended, nobody knows how best to prevent them from re-offending while protecting society from them. "Treatment" programs have ranged from primordial scream therapy to sadistic penile blockers with no decrease in offender rates. Today many "authorities" advocate lifetime tracking, public awareness, and surveillance as the only sure protection from the sex-offenders' predations.

CHAPTER TWENTY-SEVEN

AVOIDANCE THERAPY

"You are doing what?" Fettig said loudly, short of shouting. I was astonished. At our staff meeting's general discussion of Jeffrey's treatment plan I had mentioned going to the Y.M.C.A. for swimming and weight lifting which surprisingly resulted in the psychologist's emphatically negative response.

"Taking the boys to the 'Y'," I responded, hesitantly, thinking that Fettig's indignation must be the result of one of our charges exposing himself or engaging in some other deviant conduct while there. "Something wrong? Have there been complaints or reports of problems?"

"And what do you do there?" Fettig asked.

"We all go up to the weight room for twenty minutes or so then Doug takes one group swimming and I take those who want to play to the racquetball courts."

"Everybody?" he asked. "Even the S.O.'s."

"Sure," I replied.

"S.O.'s" was a fairly new acronym which still took me time to recognize. It meant youths who were designated specifically as sex offenders. I had spent almost thirty years believing that the less emphasis put on offenders' pasts the better. "A new start" and "what's past is past", I felt. But we had to designate many youths as S.O.'s because of legislation which provided funding for sex offenders at the expense of other traditional offender programs.

The sex offenders were easier to handle than most other crime-committed youth. They ran away less frequently, seldom fought among themselves, or engaged in illicit sexual activity. I had observed many of these youths to be socially misfitted, introverts, most lacking enthusiasm, hobbies, or interests except for television which they could watch days on end. Most lived in their world of private fantasies, and showed little real interest outside themselves. Fortunately S.O.'s tended to be more malleable than thieves or gangsters, and so could be coaxed—or intimidated—into at least temporarily compliant behaviors.

"Do the S.O.'s lift weights and play racquetball?" he asked.

"Yes, they have a great time. Jeffrey and I have a fierce competition going in racquetball."

"Swimming is permissible. All other competitive and machoistic activities such as weight lifting must stop immediately!" Fettig ordered.

I protested, "But I've watched our S.O.'s. They're acting more normally each time we go to the 'Y' even around the little kids . . ."

Fettig cut me off in mid-sentence, "Around what little kids? Do you mean there are small children there?"

"Yes," I replied, "A teacher I know brings in his sixth grade class to swim and so we all undress and sometimes shower together. Doug and I keep track of every student. We know what we're dealing with and we aren't going to let anything happen."

"Avoidance, avoidance, avoidance," Fettig said. "Offenders must avoid any and all contact which could cause them to revert to thoughts of abuse especially with children. Just like alcoholics avoid being around alcohol, S.O.'s must avoid deviant thoughts. S.O.'s must never again lift weights which gives them male dominance and control thoughts, compete in any sport, and never be with younger children, especially unclothed."

"But," I protested, "we watch them. Three weeks ago when the school children first joined us, our S.O.'s furtively looked at them—especially their genitals, but now less and less. I think . . ."

"It doesn't matter what you think. I am their sex offender counsellor, I say don't take them there again," Fettig stopped me.

"But if we leave them in the camp one of us must stay with them, and that sets them off even more from non-S.O.'s. They are already so dependent upon the support system they're afraid to leave. I think," I said, this time Fettig didn't interrupt me, "that if we don't change this, these inmates will be institutionalized as long as they live. They need normalcy, not just sitting around writing their fantasy memoirs in journals and reminiscing about their sexual deviancy in group sessions."

"Most should never be released. They will never be normal, no matter how normal they may seem to you," Fettig said, his voice again rising.

The Camp Director interjected himself verbally, "It's okay, Jack. Larry or Huff or I are here most 'Y' times. Just leave the S.O.'s and we'll look after them. They can do school work or write in their journals . . . or something."

"Okay," I conceded, "but I still think . . ." I stopped and looked at the people in the room. Fettig looked angry and everyone else bored. "Okay as of tomorrow this will be our policy. S.O.'s will stay at camp instead of going to the 'Y' with the group."

As it happened there was no 'Y' visit the next day because of other disruptions. We didn't go to the 'Y' for two weeks more, so by the time we did go back, common sense prevailed and I was given permission to take Jeffrey and two other S.O.'s with us provided we all changed into our gym clothes before leaving camp. We did this the full year before I retired. Everyone went to the weight room, but the heavy "macho" weights were off-limits to all. In the racquetball courts we didn't keep score so that sport was discontinued out of boredom. Everyone dressed down before leaving the camp.

Jeffrey had by then been at camp three times, and his face sheets reported that he had been in dozens of other facilities as well as lived with variety of relatives for short periods. He had run away from most of these ineptly, so as to be quickly recaptured. Like most juvenile sex offenders his crimes were mostly against friends, relatives, and family members, always younger children.

He had molested smaller children from the time he was eight years old, including several rapes of his own sister. He had killed a half-brother and shot his cousin, who had survived. Jeffrey had set dozens of fires including one at camp, and during his stay there were two other small arson fires which only by chance died out soon enough to prevent serious damage.

The larger fire was brought to my attention by a student who said he smelled smoke. At the time, Jeffrey was sitting in my classroom working on an assignment. He never looked up or showed any agitation or excitement as did other students when I left to investigate taking the youth who had reported the smoke with me.

When I opened the rest room door there was a totally black layer of smoke floating three feet above the floor to the ceiling. I took a nearby fire extinguisher while sending the boy with me to alert Mr. Hogland to call the fire department. Crouching low under the smoke I made my way to the mud room where rubberized rain gear was hanging, ablaze with an eerie orange flickering glow and the source of the inky black smoke. We were able to extinguish it with only the loss of the rain clothing and little damage to the building.

At the time I didn't suspect Jeffrey although his lack of interest in the fire should have been suspicious. His name was brought into play by another youth who claimed to have seen him putting something into a pocket of a rain coat hanging on the wall which was later ablaze.

After lengthy questioning Jeffrey admitted to smoking in that area known as the mud room. He claimed when a staff approached he had hidden a lighted cigarette in the pocket of a rain jacket. He was returned to MacLaren to one of the sex offender units there, but as usual, he wasn't there very long. Soon he was back at camp again. He had demanded a hearing and was judged to be, "Not guilty for lack of evidence."

Jeffrey was the sex offender who was able to make an excellent impression. Like many S.O.'s, he could converse on many topics with considerable awareness. His articulence belied his past record of violent, murderous conduct. Everyone who worked with him urged that he be given psychiatric treatment in a secure facility such as the state hospital unit for the emotionally disturbed, or at least at Dammasch Hospital (both since closed).

These hospitals refused to accept him because he was considered too great a threat inside their facilities. Oddly enough offenders are frequently denied treatment at mental hospitals on the basis of "being too crazy to treat" and so are left in correctional facilities from which they are eventually released to the community. Only those adjudged to be "legally insane" are hospitalized, and then should they commit criminal acts in a hospital, may be sent to secure correctional facilities for "the protection of the hospital's staff and patients." This explains, but doesn't justify, such offenders' early releases.

Our most serious incidents, one staff assault, a murder, and the highway fatalities of two inmates were the consequence of sending mentally disturbed youths directly from wards at the secure Treatment Center in Salem and Dammasch Hospital at Wilsonville. Psychologists at MacLaren faced with no place for these youths have frequently taken the easy way out and just placed them in the general population of the institution from which in a few months they are released to the communities.

Jeffrey was released to live with a relative in Washington even though he was then over eighteen. He had no marketable skills. One reason why he was so readily released was that he was being sent "out of state" in hopes that if he re-offended he would be Washington's problem to manage. Wesley Alan Dodd was handled in much the same manner.

M. Doug Hogland, science and math teacher, about 1990.

CHAPTER TWENTY-EIGHT

IN WITH THE YOUNG . . .

"Do you remember me?" our new staff member inquired.

Most of us who are older have met so many people that our memory banks become over-full, and this problem coupled with the inevitable deterioration of some memories, leaves us elders vulnerable to such embarrassing moments. I searched his features for recognition. I did remember, but I didn't, most frustrating! Recall will come, I know it will, but the immediacy of the moment makes remembrance all the more difficult.

"No, Steve, I'm sorry to say that I don't."

"1978, I was here, in your classroom—as a student."

"Oh," I said, "Do you have good memories or bad ones of that time?"

"Some of both, I suppose."

"How long were you here?"

"I think it was five, maybe six months. I was here from late summer of 1978 until early 1979."

"How'd you get sent to MacLaren in the first place?" I asked the question that many Camp alumni find difficult.

"Armed robbery," he said. "Crazy, fool thing to do. I met this older kid who I thought was 'cool' at the time—I think he's doing life for murder in the state of Alaska now—anyway we got into a lot of trouble together. I robbed a Salem bus driver at gun point, actually I used a realistic looking BB gun."

Then I did remember. I recalled that in '78 armed robbery was an unusual offense, then considered very serious since guns were a lot less available than later. I remembered how I had been given advance notice that "an armed robber" was coming to camp, but when he arrived, I found just another ordinary kid, not the desperado I had been warned of.

"Did Camp help you?" I asked.

"I guess it must have. I must have matured a lot here because after I returned home I never got into any more trouble and made a 3.8 average grade point in high school. Before I was skipping a lot of school—hardly going at all and getting really bad grades."

"So you're going to be a staff here? I should think having been here as an inmate you would know better," I quipped.

"Darold Pederson and I happened to meet, and he told me of the possibility of working here. I had just finished a four-year hitch in the Navy. My juvenile record was expunged and I was looking for work so I applied and was hired as a group life staff. I hope to finish college while I'm working here and become counsellor or possibly a parole officer."

"Well that's great!" I commented. "Welcome to the Little Top."

"The Little Top?" he said quizzically.

"Yes, many years ago I worked with a teacher named Howard Johnson. Before he was hired Johnson was asked to go to MacLaren which thereafter he called 'the Big Top', like a three ring circus he said, so much happening all at the same time. Then a week or two later while filling out an incident report on a fight he had to break up bodily, he dubbed Camp 'The Little Top'. Howard said he'd always wanted to be a wild animal trainer as a kid so he was going home to practice with a whip and chair. After that he always spoke of the 'Big Top' and the 'Little Top' rather than Camp or Campus."

"I don't recall him. I think Doug Hogland and Miss Woodman were teachers when I was here," Steve said.

"He was here in 1965, you were just a babe back then."

"How long have you taught here?" Steve inquired.

"Twenty-six maybe twenty-seven years," I replied. "I'm not really sure because last year I counted myself as fifty for two consecutive years. It's hell to get old! The mind *and* the body go."

"Why'd you stay so long? Didn't you ever think teaching more normal kids would be easier?"

"Yes, lots of times I thought of leaving, but then here I have freedom from parents and school boards who probably wouldn't understand my unorthodox methods."

"Like what?" he asked.

"Well, like a couple of years ago after first aid day, a kid who had just come in the day before, messed around all day, didn't study and so flunked his test. I scored his test and told him not to be concerned, that he'd get to take it again since it was required that he pass. He got angry, and stormed out of the room saying that he wouldn't take the 'f____' test again and I could 'shove it where the sun don't shine'. Well just outside the door he encountered Mr. Hogland who directed him physically back into my classroom. Alas, he was still belligerently vocal so I slammed him down into the chair by my desk. He sobbed that he hadn't passed because he couldn't read, that nobody ever cared enough about him to teach him to read. I felt sorry that we had been so physical with him so I assured him that. 'Mr. Hogland and I care about you and we will teach you to read', then recalling our recent violence toward him, I paused, thought a moment, and added, 'if we break every bone in your body doing it!'" We both started laughing, and he took the books and test I offered him. I'm sure that he would have passed his first aid if he hadn't been released unexpectedly two days later to go to a foster home.

"You get called on the carpet for it?" Steve asked.

"No not then, but I was twice before. Once because word had circulated that I was physically disciplining kids. That was partly true. I had never and never did hit an inmate although I had shaken several and bounced a few off the walls. When I got to MacLaren I had a conference with our Vice Principal who was pleasant as he tactfully advised me that I should 'always treat students with courtesy and consideration'. I assured him that I had done so and would continue to do so, after which we walked down a snow covered walkway to the campus galley for lunch.

From behind a bush a playful youth threw a snowball at him so he cautioned the boy to stop. After two more snowballs were thrown, he seized the boy by his shoulders and shook him like a puppy might shake a bone. As we resumed our walk, I inquired, 'What about that courtesy and consideration you told me about?' To which he replied, 'First you gotta' get their attention'."

"And what was the other time?" Steve asked.

"Well, that's always been a mystery to me. I was asked to come over to MacLaren for an important matter. When I got there, I was told that I was teaching too much grammar. I asked if I was neglecting other courses in favor of grammar because at that time I had devised a 'hangman-like' sentence diagraming game which my students enjoyed. No, I was told our students are only delinquents who don't need to know much grammar because they'll probably be truck drivers and laborers. I was amazed at what I was being told and couldn't imagine what it was all about, but I heard later that some ex-camp students were complaining about not being allowed to play our sentence diagraming games because their teachers there didn't know how to diagram sentences and they were causing problems for their teachers. While he never directly said so, I think he wanted me to stop teaching sentence diagraming with those games."

"That's hard to believe," Steve said.

Thinking he might be questioning my veracity I defended my truthfulness with an even stranger tale. "I know, but it's true. Another time I had a boy whose step-father was a biker, like a Hell's Angel, and while I never met him the boy described him well enough that I would call him a stereotypical ignorant biker slob. The kid came back from his home visit two days early, upset. His step-father had kicked him out of the house.

The boy said that he had told his mother and step-father that he had 'swum' in the Y.M.C.A. pool in Tillamook, to which his step-dad responded, 'If you're gettin' so damned uppity that you can't say "swimmed" like the rest of us, get your ass out and don't never come home again.'"

"That's too bad," commented Steve.

"Yes, his old man must have been a jerk!" I replied.

"No," said Steve, "I mean too bad that you didn't teach him 'wrong' so he could be 'right' at home with his step-father."

"Well, I'll have to think about that for a while," I said.

***

Steve Wilson continued to work at camp while he finished his college training. First he was a group life staff and later juvenile officer in charge of the Tillamook Plan in which local at-risk youths came to school at Camp either to catch up on missed studies or to wait out a semester when they were expelled. Steve has since earned his degree and is a State of Oregon parole officer working under Rick Hess who was once his PO.

In August of 1989 I retired after thirty-one years at Camp Tillamook. I was feted at a retirement party at which my bosses Richard White, Rick Hill, Bob Jester, and Principal Jim Richard said some nice things about me (re: film *Goodbye Mr. Chips*) and gave me *two* gold pens. (That retirement party had been planned for two, but Doug Hogland, my teaching partner for seventeen years had decided to hang on another year to complete his thirtieth year).

POSTSCRIPT

A FINAL FOUR

"Mary Lou has been warned that she may lose her job and along with it her insurance coverage if she misses more than three months' work. She says her contract under the Chapter One Program says that she has to teach a minimum number of days a month," my wife Fran told me. I had retired in August the year before. I knew Jerry was very ill and Mary Lou was at his bedside day and night but I didn't know she was in danger of losing her job.

"I doubt it. I've seen people off work for longer for a lot less cause," I responded. "Some for months, contemptuously thumbing their noses at Children's Services Division from their lawyers' offices with phoney 'stress' syndromes."

"Well, Mary Lou won't get a lawyer, and she won't leave Jerry's side even if it costs her job and all their savings."

"I wish I could help," my wife said.

"Me too, but I would be a very poor nurse for Jerry," I added as I left the room.

I had gone only a few steps when it came to me how I could help. I had recommended Mary Lou's hiring seven years earlier for Title One, a federally funded half-time teacher position. For many years before she was hired I had taught those students and had taken several special education classes just to learn how to teach such students, so why couldn't I work in her place? It seemed the best way for me to help Jerry and Mary Lou, and being retired, I had the time and didn't need money so I could work in her place while she continued to receive her paycheck so she could continue to qualify for her essential hospital insurance.

I picked up the phone and called Dick White and proposed that I work in place of Mary Lou, but that MacLaren School Department give her credit for my time worked. He said that he would check on my idea and call me back. He did a short time later agreeing, but stating that if I could come in as a volunteer they wouldn't pay me, but then Mary Lou could continue to sign her time sheets and receive her usual monthly check. He also said this would take the pressure off her to return to work. It seemed like a solution to her work problem, at least for a time.

The next day, Tuesday, I reported to work. Karl Morris, the teacher who had replaced me, greeted me and opened the tiny classroom we had built into one end of a larger one. I examined the available books and materials, much of which I found unfamiliar, and not very useful to my teaching style so I went to my old classroom and gathered some more familiar things.

My first student that day was Dennis. I introduced myself and asked him a few questions about himself, his home, family and what he hoped to achieve in school. As usual the answers were disappointing. He had been at MacLaren and Hillcrest for almost three years. I asked him to read a page to me, which after the first few words made it clear that he was not a Chapter One eligible student. He was reading above sixth grade level but with occasional reversals and inversions of letters so that he read "pop" as "bob".

"Do you wear glasses?" I asked him.

"No," he replied. "I don't like 'em."

"Then you have glasses?"

"Yes," he said.

"Where are they?"

"I got 'em in my locker."

"Get 'em," I said.

He left the room and returned in a few seconds with glasses in hand.

"Okay Dennis, put your glasses on and read please."

"You read very well with your glasses. Did you wear them when you took your California test?" I said.

"No, I look dumb in 'em!"

"Well everyone else thinks you're dumb when you don't wear them. Each day you come to class when I'm here, bring your glasses, Okay?"

"Now what do you want to do?"

"Finish this book that I started," he held up a thin paper backed mystery story of perhaps sixty pages.

"Okay," I said leaning back in my chair, "I'll wait for you to read it." I noted he wasn't wearing his glasses. "Please wear your glasses."

He grumbled about wearing his glasses, but did. In fifteen minutes he put the book down and said. "That was a good story!"

"Tell me about it," I replied.

He did, endlessly reciting every detail and correcting himself and referring to the book for characters' names. I tried to remain attentive and appear interested.

Finally he finished. I said, "Write a book report about it. That can be your language lesson for the day."

He said in disbelief, "You expect me to write all that!"

"No. You've got to write the story in less than one hundred words. See if you can. Okay?"

"I can't do it," he said.

"I'll show you how. Tell me again how the story starts."

After a couple minutes I stopped him. "Write all that down in one sentence with some 'because's' and 'and's', but leave out all the unnecessary stuff like saying 'Todd was a ham radio operator'."

He bent over his paper so low it became obvious that he had once again removed his glasses.

"Glasses on," I demanded.

"Oh yeah," he put on his glasses.

He summarized the story in one hundred and twenty-three words and I had to agree that was the best possible (he asked to do another book the same way and the next day turned in a ninety-three word report). Then he wanted to write poetry so I showed him a bit about it and we wrote a couple simple verses. Then my two hours with Dennis were over.

Next I talked to Rex Layton, an eighteen-year-old boy like Dennis, but with a Certificate of Attendance Diploma. Rex wore his hair in a very short Marine Corps cut. He was a good looking youth with handsomely uniform features, but pinched into the center of his head "Peanuts"-like. His bright dark eyes darted about ferret-like. In my interview he said that he was going to go live in the woods in a shack he knew about so therefore it was quite unnecessary for him to study. I asked him to read me a page from a pre-GED study book. Halfway through he tired of it and said he didn't intend to work for his G.E.D. or anything else since he was going to live in the woods on a friend's land by a lake. He would eat by catching fish and snaring rabbits. Rex told me that the world was corrupt and so society was to be shunned. He said he could best relate to nature by leading a solitary rural existence.

I told him that he didn't need to be in Chapter One classes since his tested levels were all above sixth grade, and that it simply seemed to me that he was evading his responsibility to take advantage of the camp's educational opportunities. I also commented that his idealistic image of life in the woods was neither sensible nor practical. I told him of another student whom I had taught several years before who was going to walk behind his horse tilling the soil and whose picture had appeared the previous week in the Business section of the *Oregonian* as manager of a new Portland enterprise.

This seemed to strike a note of interest. I was then sure that what he said was just rationalization not what he meant. I found him a copy of Thoreau's *Walden*, and directed him to a seat in Karl's classroom to study for his G.E.D. certificate, and if he should decide against that, Walden would describe what it's like to be a philosopher living alone in the woods by a lake. (I rather doubted that he'd be successful at either.)

My next student was Roger Sams. He stuttered very badly, particularly when he was reading and stumbled over unfamiliar words. It soon became obvious that because of his handicap and smaller size he was picked on by others in the group and was verbally belittled. He had come to camp from Hillcrest and felt that he had been tricked into leaving there. I was not surprised. He may well have been correct since kids like Sams become dependent on their Hillcrest support group and frequently are sent to camp prior to release to become more self-assured and less dependent upon "group."

He had once gone to Portland Raceway with a Hillcrest group to help out at go-kart races and had good memories and even a memento program of the events with his name printed in it. He had also rendered first aid to an injured kart driver. This made him feel good he said. I gave him a first aid book with lots of pictures to look through.

My next student was Jones. He was a short, chubby red haired boy of fourteen who was still in the seventh grade. He lisped very badly but I understood him to say that he had a gang of his own in one of the local schools. I asked if his gang were like the Crips or like the Bloods. He assured me they weren't because they didn't have guns. I said "That's good," to which he commented that they had to use knives instead and didn't allow blacks. It soon became apparent that he was very racially biased. When I asked what he was interested in, he replied, "science." A familiar book was nearby. I picked up two copies and we read a story called, "The Peanut Man" about George Washington Carver and then one about the "Professor of the Piney Woods School" in Mississippi. These were written on about sixth grade level and presented both these famous blacks in very humorous and human terms. He made no comment about the stories I had selected.

That was three weeks ago. Today Jones is gone. Bounced back to MacLaren from his day program at camp for community infractions. Layton is working on his G.E.D., still loudly condemning society and vowing that no matter how educated he becomes he is still going to live in the woods near Grant's Pass as "an armed too the teeth" survivalist hermit.

I still have Sams who has shown amazing talent with the computers. He can decipher their cryptic messages and use even unfamiliar programs. When the Image-writer printer didn't work, he took it apart, fixed it and told me how and why it worked while he was doing it. His current project is writing for a government grant to improve the park across from the camp. He is also studying first aid and hopes to become an EMT someday. I noted with interest that when he's interested in something and confident Roger doesn't stutter, or at least much less. His test scores are grade levels of 2.7 to 3.0, but today he made only two mistakes writing the draft of his grant proposal. He asked me how to spell continuity and left the "u" out of guard, although when corrected he said he knew better.

Dennis is writing. He has written two book reports and a poem about the nine lives of a cat which is excellent. He claims that it's original and yet I keep thinking I've heard it before. We may see about getting it published unless it turns out that someone else wrote it first. He is writing on the Apple Works Apple Writer which Dennis and I are both learning to use from Roger.

Jack L. Graves

DESIGNING "NO-FAIL" INSTRUCTIONAL PLANS

In the 1970's I attended several Oregon Small Schools' Conferences on individualizing instruction which proved invaluable to me since the camp school which was described as designed to educate older and more advanced students always accepted youths of greatly varying grade levels and abilities, in later years even soliciting and specializing in remediation and special education. By the '70's I found myself trying to teach up to thirty students in two dozen subject areas with as many as eighteen students in my classroom at times. To survive I was forced to prepare very specific lesson plans and attempt to orchestrate my classes like the maestro who points to the world geography elements of U.S. history, English II, III, IV, and harmonize with world history and geography. Unfortunately many a discordant note was struck in my orchestra and over my baton I could see that some of my musicians weren't playing at all, much less harmonizing.

What I needed was time. Time to tutor, time to lecture, time to prepare. Individualization was the answer. By using ideas picked up from the Oregon Small Schools workshops, combined with others presented by some special educators of that time, such as William Glasser (in his 1958 *Schools Without Failure*), I devised my own no-fail system which served me well for many years. I modified Glasser's and others plans and expanded the curriculum far beyond what they proposed since they were concerned more often with remedial education. My plan had to have something for everyone: in fact, I found it was well accepted and even more successful with academically advanced students because it freed them to learn as much, as fast, and in some cases, what they wanted.

A NO-FAIL PLAN OF INDIVIDUALIZED INSTRUCTION

This plan is adaptable to any school system (as large as a whole state, or as small as a single classroom). It allows for individual differences, permits open entry and exit, may be credit or non-credit, graded or ungraded. It frees teachers to spend more time with those students who need them most, and enables those who do not to forge ahead without being singled out and identified as "hares, fast trackers, or TAG's"—all labels which may place unnecessary social or academic pressures on classmates. Ambition may be rewarded without diminishing the self-esteem of slower, but conscientious class members. Those students who would otherwise become dropouts and disrupters will find themselves less able to deny responsibility for their lack of achievement by excusing their non-performance with the usual "I can'ts, I didn't understand", or "the teacher didn't have time to help me" excuses.

For example, "Communications" could perhaps be split into areas such as writing, reading, and speaking, and then limited and defined to achieve sufficiently specialized course designs (I used a design of eight basics such as sentences, parts of speech, spelling, letter writing, etc., supplemented with advanced courses such as creative writing, writing the research paper, and a variety of special interest literature classes).

The process of designing a course could result in a final course description something like: Communications IV—reading and writing: "Appreciating and Creating Poetry": a 1/4 credit course. (Credits and grades are optional to the plan but may serve as incentives, rewards and/or measurements of achievement to report to the student, school and/or parents.)

If this process sounds familiar to those of you who have been involved in Oregon education over the past twenty or so years, I present no apologies for I hope the plan incorporates the very best of what, I believe, has transpired (and frequently expired) in educational design in this state.

Previous meritorious models have failed because they became too formalized, involved and were unnecessarily complicated. To succeed any plan must be functional, versatile, and simple. The key word is "simple", which means readily understood by teachers, students, parents, even administrators, and easily implemented so that neither the planners nor the users lose sight of their instructional objectives. Some years ago in Oregon, "goals and objectives" were stylized into gibberish. A jargon that became an "end" rather than "a means to an end" so that "educators" spent most of their time and energies defining and refining goals and objectives, but few implemented them in their classrooms.

The much vaunted Japanese schools rely on central direction from a national department of education, which unlike ours, directs every detail of curricula content so that on any given day all schools in Japan are teaching the same subject matter in much the same way. In Japan upper grade classes are large, and teachers may lecture to 50 or 60 students in a formal manner. The advantage of this is that teachers do not have so many class preparations and students do not have as many classes in a day or a week, but rather have "study time" in which they are available in their offices and classrooms to tutor or be tutored and to work on what in our system euphoniously referred to as homework. (In this no-fail plan, homework is whatever the student feels she or he wants to do, or should do to attain their individual goals (personal or assigned), beyond that there is no homework).

The U.S. student is generally less socially obedient and so less likely to sit for 60 to 90 minutes listening to a lecture than Japanese students so instead of lecture periods this plan incorporates lesson "contracts" which outline the requirements in the form of reading, written and oral recitation, and individual instruction, necessary to attain competency in each course and so be eligible to move on (or to earn a given unit of credit). Of course competency (and/or credit) eligibility must be ascertained by some means such as traditional testing, conferences, or as in physical or manual arts, demonstrations of present or evidence of past performance sufficient to indicate demonstrable knowledge to be able to repeat that skill or performances on demand (such as welding, writing, music, etc.).

I refer to this plan as "no fail" because (if grades are used) it is never necessary to fail any student for any reason. Those students who do not demonstrate themselves competent, continue to prepare until they are able to do so. So that in the example earlier I chose to use, "COMMUNICATION, Appreciating and Writing Poetry", the design committee would also design related supplemental projects for those who need more time and study to attain the curriculum goals presented in the "course outline" or "master sheets" (or whatever the printed course competency requirement sheets may be called). Students are thus enabled to progress at their own pace, choose specific areas of interest, begin and complete courses at differing times, and are denied the competitive "I can't do it because it's too difficult" excuse. They may be given easier projects and or more study time and/or tutoring. Everybody succeeds—but some take longer than others—thus students are able to achieve at their own pace to attain pre-determined demonstrable levels of achievement.

Courses may be designed at any level from middle school through graduate classes, but I believe are best suited to high school where students are demanding more independence and individually challenging course content (and are also demanding to be "treated as adults"). Such "self directed" courses of study may be designed for remedial, vocational, arts, and technical, as well as for high level academic learners.

Teachers become more important than ever in this plan, but as mentors, not lecturers as in Japan or the university, but rather as enablers, tutors, and counsellors. They will have more freedom to make decisions as to how to allocate their time, and may work as individuals, teams, or even on computer disks or VCR recordings. The possibilities are endless for imaginative instruction, but should not be isolated with certain kinds of learner groups. When a teacher finds him/herself working with students outside his expertise he should not despair but instead think of himself as a "learning facilitator" and so refer such students to other teachers, community resources, persons, or library sources, to provide those students with the best possible instruction in that particular discipline or subject area.

A threat to the success of this kind of plan lies in permitting snobbery or promoting the special education "turtles and hares" syndrome. It is essential that all students be treated as "equal learners" no matter what their levels or courses of study, and that only confidential individual notice be taken of students' successes or lack of them (there is no failure!).

The most difficult process in American education is determining not how to teach but rather what to teach. This is often nearly impossible because of ever-changing ideas by those who "lead" the American school systems. What to teach or what not to teach is often determined politically by those whose ideologies, philosophies, or funding may be impacted. Fortunately funding requirements to implement these plans may be either lavish or spartan—if spartan, students may be steered to off-campus sorts of learning activities and sources to compensate for the lack of those in their school(s).

All of the above are prerequisite to any educational design, but are not factors in an educational plan unless as in Oregon at this time, there is no established consensus about curriculum in which case "STEP 1" must designate what to teach. As Oregonians today we will probably have to establish what to teach before we determine how to teach. Group planning sessions should be designed as appropriate to the local school structure. The larger districts must necessarily divide their planning sessions, in the interest of efficient participant input, by discipline, department, interest, and public prescription.

***

STEPS TO DESIGN AND IMPLEMENTATION

STEP 1: WHAT TO TEACH: In most situations a curriculum is pre-determined by the state educational authority. These directives then become "objectives" and "curriculum (course goals to students) goals" but may need to be further defined by participants in the planning process so that there is general support and agreement concerning directions. Since most school districts', and states' standards are inadequate to "flesh out" an adequate curriculum, it seems best to err on the side of exceeding such standards rather than just meeting them.

STEP 2: PREPARING COURSE OUTLINES: This can become so involved and controversial that committees should probably be limited to specific courses and probably should not exceed half a dozen members. This type of organization also should prevent this task from becoming an "end" instead of a "mean." It should also be noted that course outlines are always subject to revision and that those which do not meet student approval or need must be immediately rewritten and/or revised to correct their inadequacies.

STEP 3: ORIENTATION: This simply means acquainting all participants: students, teachers, administrators, and parents, with the system and how it is intended to function. This shouldn't be very difficult for if the preparation stage has succeeded the entire program is simple to understand and easy to follow. If it's not, fortunately it will fail at this point (as the much misunderstood, loosely defined Oregon Plan of Education for the Twenty-first Century already has).

Such plans (if broadly enough written and implemented with diligence and common sense will draw few of the usual complaints (too expensive, lacking in basics, too much emphasis on one groups' expectations at the expense of another's, etc.) for it will offer something relevant to everyone and most groups.

STEP 4: IMPLEMENTATION: No program can succeed unless it is readily understood and supported by most of those using it. This system starts with determining each student's need based on: (1) what he or she wants to do, (2) what must be done to meet state testing, and/or graduation requirements, as well as what pre-tests, IEP's (individual educational plans) determine as necessary, and (3) to satisfy the individual's personal achievement goals.

In nearly every case, if the planning committees have done their work well, there will be a pre-planned course of study adaptable to each student's needs. Student(s) should be able to modify existing plans or be encouraged to devise appropriate new ones. Standards to achieve competency throughout the scope of the plan must be upheld; however, always attainable and sufficiently challenging when matched to the ability level of each student who is enrolled. Course outlines must be prepared in a simple straightforward manner with the various assignments, projects, and other learning activities required designed to "encourage" not to "test" while determining competencies. In many instances it may be advisable to break longer curriculum elements into smaller, more attainable units so that students do not become overwhelmed and are to better recognize and measure their progress as steps toward clearly defined goals: however, such goals may be presented as indicators of achievement. There must also be some provision for supplemental kinds of learning strategies to provide for those students who will require additional instruction and/or learning activities to achieve demonstrable competencies or test levels.

The following are sample pages prepared for an individualized no fail plan at Camp Tillamook (I used from 1973 until I retired in 1989). This plan was constantly modified and up-dated to meet changing needs. For most of those years it was well received by students and proved quite successful in helping most pass their G.E.D.'s or succeed in public schools after their releases.

After about 1985 changing policies and philosophies did not permit long range planning and goal setting, so I combined "No-Fail" with computer tutorials written for Tandy TRS computers. The first three specimen sheets that follow are student information sheets and the last three are course (assignments) contracts.

EXAMPLE PAGE

Camp Tillamook is a facility for court-committed boys operated under the administration of MacLaren School for Boys. The Camp, which is located on the Oregon coast two miles south of Tillamook, has a maximum capacity of 29 boys. This camp has been in operation for over 29 years serving the needs of over 5000 boys.

The philosophy behind the camp program is "work-study". The residents of Camp Tillamook are in the classroom half their time and working the other half. Their work is performed for the Oregon State Highway (Parks) Division. It is hoped that through well-supervised work instruction the student will learn skills, basic work habits, and attitudes that may contribute to his feeling of self-esteem and confidence thus enabling him to become a happier and more productive family leader. At the same time, by attending school the student may maintain status for re-entry into public school.

Camp Tillamook currently has 15 staff members. Personnel at the Camp are headed by a Camp Director and his Assistant. The staff is comprised of: two cooks, one counselor, three teachers, and six G.L.C. line staff. The Camp is housed in a seven thousand square foot multi-use building and a small storage shed and shop.

The State Parks provides two work supervisors who are responsible for supervision and training of the boys placed in their charge. The work crews engage in park maintenance, trail building and other tasks assigned by the district parks manager. The work day is basically 8:15 AM to 4:00 PM. Transportation is provided by the parks. The days work may be as far north as Astoria or as far south as Fogarty Creek State Park (about 190 miles apart). When a student-worker arrives in Camp he is palced on entry-level status for less than one week, then later he may progress through worker 1, 2, 3 and on to leadworker; with each promotion he received pay increments to a top of $3.50 per hour. In the Camp he will receive daily "grades" which reflect his performance and attitude in the Camp and school. These grades are averaged each week to place him on the appropriate level (privilege status). Then based upon his "level" he will participate in activities with various degrees of supervision. Important to the Camp concept is community contact and accordingly the youths at the Camp will go into the Community for various activities, perhaps shopping, movies, bowling, public service projects, as examples.

The Camp school program enrolls students grades seven through twelve. In addition special programs work with remedials and equivalency preparation, etc. All instruction is individualized utilizing learning packages and special materials selected for the attainment of specific goals. The school is a unit of Lord High School, MacLaren School for Boys, and, as such, has access to files and testing information from the parent institution. In addition the Camp is assisted by the local educational service district and local school district which enables the Camp to have access to audio-visual supplies and services and duplicating and testing services. As each boy enters the Camp (most are referred from MacLaren), his educational history and needs are analyzed in conference with him. The goals established at this conference are coordinated through the Camp Counselor and parole counselor to reflect his projected relase date and post parole planning. The Camp school, as a unit of MacLaren, is fully accredited and work in the Camp school is transferrable throughout the State. Nearly all students who complete their program at Camp achieve their goals, most are re-enrolled in public schools, and each year about 20 complete their successful testing and are awarded equivalency certificates.

EXAMPLE PAGE

CLASSES AND PROGRAMS AVAILABLE AT CAMP TILLAMOOK

GRADUATION REQUIREMENTS: 23 credits as follows: 10 elective, 4 language arts, 2 math, 2 social studies/history, 1 economics/government, 2 science, 1 health ed., 1 physical ed. 1 personal finance, ½ career ed. Those above plus 1 in United States history and ½ in world history or georgraphy, also the 4 communications credits must include 1 in writing/composition.

The courses below marked "R" are required now (but not necessarily all so marked). Those marked "sp" are special need offerings. Those marked "E" are elective that may count toward graduation.

COMMUNICATIONS/LANGUAGE ARTS

R 7th and 8th language arts.....G.S. cr.
(grammar, spelling, reading)

R high school communications basics
(basic prescriptive instruction in 10 areas based on pre-test diagnosis)

R high school writing
(sentence, paragraph, composition)

R High school literature courses
- The Short Story............½ cr.
- Literature Survey I & II...1 cr.
- Lit. and Cr. Writing.......½ cr.
- A Shakespeare Play.........½ cr.
- Analysis of Poetry.........½ cr.

SOCIAL SCIENCES

R 7th and 8th grade social studies..G.S. cr.

E Global studies....................1 cr.
E World Geography...................1 cr.
E Social Responsibility.............1 cr.
R Consumer Education................1 cr.
R Career Education..................2 cr.
E Economics I & II..................1 cr.
E Advanced Economics................½ cr.
E World History.....................1 cr.
R United States History.............1 cr.
R Government (2 courses)............1 cr.
E Modern Problems...................1 cr.
E Black History.....................½ cr.
E Juvenile Law......................½ cr.
E Map Reading and Making............½ cr.

REQUIRED COURSES (each ½ credit per sem.)
All students must enroll in the following and successfully complete:

R Health and First Aid (MM I card required)
R Physical Education ($3.00 fee required)

E MINI-COURSES (each possible ½ cr.)
Art, Aviation Theory, Basic Auto, Sport Fishing, Driver's Ed. (manual only), others may be devised as appropriate.

SCIENCE AND MATH

R-E General Math.........1 cr.
R-E Consumer Math........1 cr.
R-E Personal Math........1 cr.
E Algebra..............1 cr.
E Geometry.............1 cr.

R-E General Science......1 cr.
R-E General Biology......1 cr.
R-E Human Biology........1 cr.
E General Chem. (dry)..1 cr.

GENERAL EDUCATIONAL DEVELOPMENT (GED)

sp. Preparation for the five tests as follows: Reading, Writing, Social Studies, Science, and Mathematics.

Tests administered locally. There is a $20.00 fee for testing.

No credits awarded for GED Prep.

C.L.E.P. (College Credit by Examination)

sp. Preparation for those who have earned a GED or have graduated. This program enables those who wish to prepare for and pass tests at the college level. A $35.00 fee is charged for the actual testing, none is charged for preparation.

TITLE I (BASIC SKILL IMPROVEMENT)

sp. This will help students to bring up their levels in reading, math and language arts..This may in some instances be a program requirement at Camp. Credit may be earned upon completion by testing to prove competency at grade level in basic skills.

DIPLOMAS AND CERTIFICATES OF COMPETENCY

It is possible to issue diplomas, certificates of attendance, and competencies.

ALSO NOTE: Work crew is a credit class. Each semester of work experience counts 1 credi

EXAMPLE PAGE

```
                         LESSONS ON COMPUTER DISK

AUTHOR PROGRAM LESSONS

FIRST AID: MMFA/LSN     REVIEW OF MULTI-MEDIA FIRST AID       50 FR/13 GRNS
           MMFA2/LSN    ADVANCED FIRST AID PROBLEMS           30 FR/17 GRNS

AUTOMOBILES: ORDRPREP/LSN     1/3 OF OREGON DRIVERS MANUAL     52 FR/30 GRNS
             ORDEFDR/LSN      1/3 OF OREGON DRIVERS MANUAL     40 FR/30 GRNS
             ORDRAUTO/LSN     1/3 OF OREGON DRIVERS MANUAL     35 FR/17 GRNS
             ORDRDEF/LSN      DEFINITIONS FROM OR. DR. MAN.     5 FR/5  GRNS
             CARBUY1/LSN      LESSONS ON FINANCING CARS        MOST 10  QUEST.
             CARBUY2/LSN         AND SELECTING THE RIGHT
             CARBUY3/LSN         CAR BASED ON NEED AND
             CARBUY4/LSN         ABILITY TO PAY.......          EACH 17 GRNS

LEGAL RIGHTS: ORJUVLAW/LSN    METROPOLITAN YOUTH COMMISSION     40FR/33 GRNS

G.E.D. PREP.:AMCOLHST/LSN      COLONIAL HISTORY                 50 FR/38 GRNS
             AMHIST2/LSN       U.S. HISTORY 1800-1900           43 FR/32 GRNS
             AMHSTNOW/LSN      U.S. HISTORY 1900-NOW            47 FR/33 GRNS
             CONSTIT/LSN       U.S. CONSTITUTION                35 FR/27 GRNS
             GEDECON1/LSN      BASIC ECONOMIC TERMS             42 FR/23 GRNS
             GEDECON2/LSN      ECONOMIC SYSTEMS                 44 FR/24 GRNS
             GEDLIT1/LSN       LITERATURE (HISTORY OF)          59 FR/37 GRNS
             GEDLIT2/LSN       LITERATURE (INTERPRETATION       64 FR/43 GRNS
             GEDLIT3/LSN       PRACTICAL READING                21 FR/11 GRNS
             WRISKIL1/LSN      SPELLING  AND MECH.              52 FR/27 GRNS
             GEDUSA1/LSN       G.E.D. USAGE PRACTICE            42 FR/27 GRNS
             GEDMECH/LSN       SENT. STRUCT. AND MECH.          43 FR/24 GRNS
             GEDSTYLE/LSN      STYLE PRACTICE                   19 FR/11 GRNS

MISCELLANEOUS: BLETWRIT/LSN    BUSINESS LETTER FORM             17 FR/11 GRNS
-----------------------------------------------------------------------------

QUICK QUIZ TESTS WRITTEN
 READ 2,4,6; COMP 2,4,6,; WORD 2,4,6,; SPELL 2,4,6...   CH 1 READING DIAGNOS.
 MMFAT1, MMFAT2, MMFAT3, MMFAT4.........................MULTI-MEDIA FIRST AID
 SPELL40W AND SPEL40W2............................HIGH SCHOOL LEVEL SPELLING
 ORJUVLAW..........................................OREGON JUVENILE LAW TESTS
 ORDRTEST..........................................OREGON DRIVERS PRACT.
 COMPUTES AND COMTRS80.............................COMPUTER LITERACY TESTS
 PARTOFSP, MODPHCL, VERBTEST, KINDSENT...............ENGLISH GRAMMAR TESTS
 CONSTIT1.........................................TEST OVER THE CONSTITUTION
```

EXAMPLE PAGE

NAME______________________________ PRE-TEST: ENGLISH BASICS GRADES 9-11

....

INFORMATION: The purpose of this test is to find out what you already know in English and language arts. You will then be able to concentrate on those areas which you most need to assist you in the future.

....

JUST DO THE BEST THAT YOU CAN AND AS MUCH AS YOU CAN...THIS IS IMPORTANT TO YOU!

....

BASICS I: PARTS OF SPEECH: There are eight parts of speech which are determined by the way a word is used in a sentence- that is, what it does in the sentence. Naming words are nouns. Those that take the place of names are pronouns. Words that show action or state-of-being are verbs. If a word joins words or parts of the sentence, it is a conjuction. Those words that modify or describe are either adjectives or adverbs. Words that exclaim are interjections, and those that relate other words (nouns or pronouns) to the sentence are prepositions.

....

INSTRUCTIONS: In the sentence below there are ten numbered, underlined words. Determine what part of speech each is and write that part of speech on the line that follows it. REMEMBER THE PARTS OF SPEECH ARE THOSE UNDERLINED WORDS IN THE PARAGRAPH ABOVE.

SENTENCE: Harry[1] told[2] us[3] about[4] his[5] very[6] exciting[7] adventure[8] last[9] summer[10].

1. Harry______________________ 6 very______________________
2 told______________________ 7 exciting______________________
3 us______________________ 8 adventure______________________ 7 correct is excellent
4 about______________________ 9 last______________________ 5 correct is good
5 his______________________ 10 summer______________________

....

BASICS II: SENTENCE STRUCTURE

INSTRUCTIONS: In the sentence below, there is a subject, a verb, and a direct object complement. Select the words in the sentence that are these parts and write them on the lines provided.

SENTENCE: Before the jump, Louis packed his parachute in the chute pack.

SUBJECT__________________ VERB__________________ COMPLEMENT__________________

....

BASICS III: PARAGRAPHING: This is the most important part of this pre-test. It will show whether or not you can organize your ideas and express them clearly. If you do nothing else on this test, complete this part the best that you can. Also use your best penmanship, spelling, and punctuation. Write your paragraph on the lines provided at the top of the next p ge. The topic you should write on is your first impression of Camp Tillamook.

....

CONTINUED ON THE NEXT PAGE..............

EXAMPLE PAGE

U.S. HISTORY - SECOND SEMESTER R EQUIREMENTS

SOURCE BOOK - AMERICAN NATION - CHAPTER 17 - 41

PROCEDURE: STEPS: (1) READ EACH CHAPTER (2) DO THE ASSIGNMENT & TURN IN
(3) TAKE CHAPTER TEST, IF PASSED GO ON TO NEXT CHAPTER

CHAPTER 17 - P.333-345
PAGE S.S.QUEST
337 2
340 2-3-5
343 2-3
345 none
ALSO: THOREAU
HORACE MANN
FUGITIVE SLAVE LAW
ABOLITIONIST

CHAPTER 18 - P.350-366
PAGE S.S.QUEST
353 1-2
356 1-4-5
361 2-3-4
363 1-2-4
366 1-3-4
ALSO: FT. SUMTER
KANSAS-NEBRASKA ACT
POPULAR SOVEREIGNTY
APRIL 12, 1861

CHAPTER 19 - P.368-369
PAGE S.S. QUEST.
372 1-2-4
379 1-3-4
383 2-3-4-5
386 1-2
389 1-3
ALSO: U.S. GRANT
ROBERT E. LEE
ANDREW JOHNSON
"STONEWALL" JACKSON
APRIL 9, 1865
FORD THEATER
GETTYSBURG

MAP AND COMPARISON CHART
OF NORTH & SOUTH
GET FROM TEACHER

TO TOP OF NEXT COLUMN

CHAPTER 20 - P.395-413
PAGE S.S. QUEST.
399 3-4
402 3-4
405 2-3-4-5
408 2-5
413 1-3-5
ALSO: SOLID SOUTH
RECONSTRUCTION
SUMNER
STEVENS

SKIP CHAPTER 21

CHAPTER 22 - P.430-445
PAGE S.S. QUEST.
435 3-4
438 1-2-4
442 1-4
445 4
ALSO: 1890
LAST FRONTIER
RESERVATIONS
GHOST TOWNS

SKIP TO CHAPT. 30
CHAPTER 30 - P 561-576
PAGE S.S. QUEST
564 1-2
569 1-2-4-5
572 1-3
576 1-2-4

CHAPTER 31 - SKIP
CHAPTER 32 - P.592-613
PAGE S.S.QUEST.
595 NONE
598 1-2-3-4
602 1-2-5
606 1-3-4
611 1-3-4-5

CHAPT. 32 CONTINUED
TOP OF NEXT COLUMN

CHAPTER 32 - ASS. CONT.
ALSO: 1914 WILSON
APRIL 6, 1917 PERSHING
NOV. 11, 1918 FOCH

CHAPTER 33 - P.616-629
PAGE S.S QUEST
620 1-5
625 2-4
629 3-4-5
ALSO OCT. 1929
DEPRESSION
SKIP TO CHAPT. 39
CHAPTER 39 - P 717-739
PAGE S.S. QUEST.
724 1-2-3-4
728 1-2-3-5
734 1-2-3-4
739 1-4
ALSO: STALIN DEC.7,
DEGAUL 1941
McARTHUR June 6,
TRUMAN 1944
ROOSEVELT Aug. 6,
V-E DAY 1945
V-J DAY Sept. 2,
1945

SKIP TO CHAPT. 41
CHAPTER 41 - P.765-794
PAGE S.S. QUEST
769 3-4-5
772 3-5
777 1-2
780 1
783 NONE
792 3-4-5
ALSO: VIET NAM
VIET CONG
CIA
OAS

FINAL EXAM

EXAMPLE PAGE

ADVANCED LITERATURE/SOURCE <u>INSIGHTS INTO LITERATURE</u>
CLASSIC PROSE AND POETRY AND CREATIVE WRITING

SECTION I - Forms, devices and style: Define the Terms Below:

1. prose
2. poetry
3. short story
4. novel
5. drama
6. narrative poetry
7. lyric poetry
8. epic
9. satire
10. blank verse
11. tree verse
12. rhyme
13. rhyme scheme
14. plot
15. point-of-view
16. climax
17. setting
18. alliteration
19. meter
20. simile
21. metaphor
22. personification

SECTION II - Authors - Tell what each writer below contributed to our literature:

1. Geofrey Chaucer
2. William Shakespeare
3. Edgar Allen Poe
4. Nathaniel Hawthorne
5. T. S. Eliot
6. E.E. Cummings
7. Stephen Crane
8. Herman Melville
9. Mark Twain
10. Lory Byron

SECTION III - Read each of the following and write a brief explanation of each one's form, style and development by use of devices in Section I:

"An Occurence at Owl Creek Bridge" - p.145
"On the Road" . . . . - p.254
"The Necklace" . . . . - p.25
"The Destruction of Sennacherib". . - p.186
"Sea Fever" .... - p.189
"The Express"..................... - p.378
"The Bells"....................... - p.379
"The Snake"....................... - p.194
"The Brute"....................... - p.489

SECTION IV - WRITING (use Warriner's Orange Book)

Read p. 458 to 460 - Do Ex. 1 or 2 on P. 460
Read p. 461 to 467 - Do Ex. 4 on p. 467
Read p. 468 to p. 477 - Do Ex. 10 on p. 477-478
Read p. 478 to P. 480 - Do Ex. 14 on p. 480

Plan and submit one of the following:

(A) Five original short poems or fewer longer ones
(B) One short story (about 1000 words)